AF749599

Nanocrystalline Materials

Edited by Behrooz Movahedi

Published in London, United Kingdom

IntechOpen

Supporting open minds since 2005

Nanocrystalline Materials
http://dx.doi.org/10.5772/intechopen.78515
Edited by Behrooz Movahedi

Contributors
Behrooz Movahedi, Luis Castañeda, Vladimir Volodin, Vladimir Gritsenko, Andrei Gismatulin, Albert Chin, R.A. Ilyas, S.M. Sapuan, M.S.N. Atikah, Rushdan Ibrahim, A. Atiqah, M.N.F. Norrrahim, M.N.M. Ansari, Hiroaki Matsui, Malik Anjelh Baqiya, Retno Asih, Muhammad Gufron, Dwi Yuli Retnowati, T. Triwikantoro, D Darminto, M Mastuki

© The Editor(s) and the Author(s) 2020
The rights of the editor(s) and the author(s) have been asserted in accordance with the Copyright, Designs and Patents Act 1988. All rights to the book as a whole are reserved by INTECHOPEN LIMITED. The book as a whole (compilation) cannot be reproduced, distributed or used for commercial or non-commercial purposes without INTECHOPEN LIMITED's written permission. Enquiries concerning the use of the book should be directed to INTECHOPEN LIMITED rights and permissions department (permissions@intechopen.com).
Violations are liable to prosecution under the governing Copyright Law.

CC BY

Individual chapters of this publication are distributed under the terms of the Creative Commons Attribution 3.0 Unported License which permits commercial use, distribution and reproduction of the individual chapters, provided the original author(s) and source publication are appropriately acknowledged. If so indicated, certain images may not be included under the Creative Commons license. In such cases users will need to obtain permission from the license holder to reproduce the material. More details and guidelines concerning content reuse and adaptation can be found at http://www.intechopen.com/copyright-policy.html.

Notice
Statements and opinions expressed in the chapters are these of the individual contributors and not necessarily those of the editors or publisher. No responsibility is accepted for the accuracy of information contained in the published chapters. The publisher assumes no responsibility for any damage or injury to persons or property arising out of the use of any materials, instructions, methods or ideas contained in the book.

First published in London, United Kingdom, 2020 by IntechOpen
IntechOpen is the global imprint of INTECHOPEN LIMITED, registered in England and Wales, registration number: 11086078, 7th floor, 10 Lower Thames Street, London, EC3R 6AF, United Kingdom
Printed in Croatia

British Library Cataloguing-in-Publication Data
A catalogue record for this book is available from the British Library

Additional hard and PDF copies can be obtained from orders@intechopen.com

Nanocrystalline Materials
Edited by Behrooz Movahedi
p. cm.
Print ISBN 978-1-78985-593-7
Online ISBN 978-1-78985-594-4
eBook (PDF) ISBN 978-1-83880-274-5

For EU product safety concerns:
IN TECH d.o.o., Prolaz Marije Krucifikse Kozulić 3, 51000 Rijeka, Croatia, info@intechopen.com or visit our website at intechopen.com.

We are IntechOpen,
the world's leading publisher of
Open Access books
Built by scientists, for scientists

4,600+
Open access books available

119,000+
International authors and editors

135M+
Downloads

151
Countries delivered to

Our authors are among the
Top 1%
most cited scientists

12.2%
Contributors from top 500 universities

WEB OF SCIENCE™

Selection of our books indexed in the Book Citation Index
in Web of Science™ Core Collection (BKCI)

Interested in publishing with us?
Contact book.department@intechopen.com

Numbers displayed above are based on latest data collected.
For more information visit www.intechopen.com

Meet the editor

Dr. Behrooz Movahedi obtained his Ph.D degree in Materials Engineering at the Isfahan University of Technology in Iran in 2010. During this period, while on a sabbatical leave, he visited the School of Materials Science and Engineering in Nanyang Technological University in Singapore. After that he joined the Department of Nanotechnology Engineering at the University of Isfahan. Recently, he became an Associate Professor of Nanotechnology in the Faculty of Advanced Sciences and Technologies. Dr. Behrooz Movahedi has over 10 years of experience in nanotechnology, amorphous materials, optical ceramics, and advanced thermal spray coatings for environmental and industrial applications. He has published three books, one patent, and more than 70 papers on nanomaterials. He was invited as a reviewer for several potential ISI journals such as *Materials & Design*, *Journal of Alloys and Compounds*, *Surface and Coatings Technology*, *Applied Surface Science*, *Journal of Materials Engineering and Performance*, *Materials Science & Engineering B*, and *Ceramics International*.

Contents

Preface XI

Section 1
Introduction 1

Chapter 1 3
Introductory Chapter: Nanocrystalline Materials
by Behrooz Movahedi

Section 2
Nanocrystalline Ceramics 7

Chapter 2 9
Silicon Nanocrystals and Amorphous Nanoclusters in SiO_x and SiN_x: Atomic, Electronic Structure, and Memristor Effects
by Vladimir Volodin, Vladimir Gritsenko, Andrei Gismatulin and Albert Chin

Chapter 3 33
Ferrite-Based Nanoparticles Synthesized from Natural Iron Sand as the Fe^{3+} Ion Source
by Malik Anjelh Baqiya, Retno Asih, Muhammad Ghufron, Mastuki, Dwi Yuli Retnowati, Triwikantoro and Darminto

Chapter 4 55
Surface Plasmons in Oxide Semiconductor Nanoparticles: Effect of Size and Carrier Density
by Hiroaki Matsui

Section 3
Nano Biomaterials 73

Chapter 5 75
A Facile Method for Formulation of Atenolol Nanocrystal Drug with Enhanced Bioavailability
by Luis Castañeda

Chapter 6 89
Production, Processes and Modification of Nanocrystalline Cellulose from Agro-Waste: A Review
by R.A. Ilyas, S.M. Sapuan, R. Ibrahim, M.S.N. Atikah, A. Atiqah, M.N.M. Ansari and M.N.F. Norrrahim

Preface

Nanocrystalline materials have been attracting much attention in the field of future science and technology. Indeed, the scientific interest in these materials in the powdered and bulk form is associated mainly with expectations of various size effects on their properties. Thus, the term "*nanocrystalline materials*" relates to the sizes of structural elements. The range of application of these materials is huge, such as more efficient catalysts, films, magnetic materials, protective coatings, and biological and biomaterials. Many compounds and elements, if made in the nanoscale, behave quite differently from how they would have in their conventional state, and these new physical and chemical properties give rise to a whole new field of scientific study.

The overall purpose of this book, "*Nanocrystalline Materials*", is to present selected advanced topics on nanocrystals, allowing the book to be a good resource for scholars and students of material science, nanotechnology and physical chemistry. The chapters for this book have been contributed by the most respected researchers in this area and give an overview of selected properties and applications of nanocrystalline materials. IntechOpen and myself as the editor hope that this book will aid current research and prove to be very useful to the scientific community. If so, this will be the nicest reward for us. Ultimately, I wish to thank all the authors for their contributions and I should like to acknowledge the sustained helpfulness and dedication of the publisher's staff, in particular of *Mr. Gordan Tot* by his insistence for this interesting project.

Behrooz Movahedi
Associate Professor,
Department of Nanotechnology Engineering,
Faculty of Advanced Sciences and Technologies,
University of Isfahan,
Isfahan, Iran

Section 1

Introduction

Chapter 1

Introductory Chapter: Nanocrystalline Materials

Behrooz Movahedi

1. Introduction

Nanocrystalline materials have been a hot research topic over the past 30 years. These materials abound in industry, bridging the gap between molecular and macroscale objects. Nanocrystalline materials are ultrafine-grained single-phase or multiphase polycrystals with grain sizes in the range of 1–100 nm, as depicted in **Figure 1**, the transmission electron microscopy (TEM) image of Fe-based nanocrystalline coating. In fact, the extremely small sizes and a large volume fraction of the atoms are located at the grain boundaries; on the other hand, these materials consist of about 50 vol.% crystalline component and 50 vol.% interfacial component.

It is recognizable that the nanocrystals are typically specified as anything, such as small grain polycrystalline materials, nanosynthesized surfaces, nanoparticles, and polymer micelles; each of them has varied usages, from drug delivery, to super capacitors, catalysts, and sensors. These materials are of interest for the following reasons:

1. The properties of nanocrystalline materials differ from the properties of single crystals and coarse-grained polycrystals and are amorphous with the same chemical composition. This deviation is strongly related to the reduced crystallites size as well as the large amount of grain boundaries between adjacent crystallites.

2. The concept of nanocrystalline materials seems to authorize the alloying of components which are immiscible in the solid or molten state. These fabricated alloys could be good candidates for advanced and technologically marvelous properties.

It is clearly seen that at a nanometric scale, the nanocrystalline materials contain a high grain boundary volume fraction; therefore grain boundaries and their interactions with crystal play a remarkable role in the different properties. It is important to point out that those nanocrystalline materials, as a new generation of advanced materials, have superior properties to conventional coarse-grained polycrystalline materials. They exhibit outstanding mechanical and physical properties such as high strength and hardness, low elastic modulus, improved ductility/toughness, excellent fatigue and wear resistance, increased diffusivity, higher electrical resistivity, reduced density, higher thermal expansion coefficient, enhanced specific heat, lower thermal conductivity, and better soft magnetic properties.

Nanocrystalline materials can be fabricated by gas condensation, plasma deposition, spray conversion technique, mechanical alloying, and some other methods.

IntechOpen

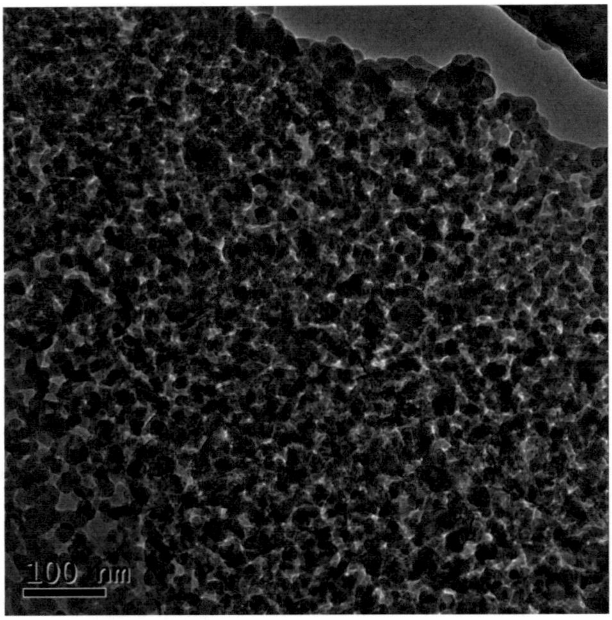

Figure 1.
TEM image of Fe-based nanocrystalline coating (unpublished image).

Obviously, there are two approaches to fabricate the nanocrystalline materials: "top-down" and "bottom-up." Both approaches play significant roles in industry and have some advantages and disadvantages. Bottom-up approach is nothing new in material synthesis and often emphasized in nanotechnology literature. As a matter of fact, the typical synthesis of materials is to build atom by atom on a large scale and has been used for over a century in industrial applications. Bottom-up approach mentions the buildup of a material from the bottom as molecule by molecule, atom by atom, or cluster by cluster. In the process of crystal growth, the growth species such as atoms, molecules, and ions assemble into crystal structure one after another after impinging onto the growth surface. Bottom-up approach also promises a preferable chance to get nanocrystalline materials with less defects, more homogeneous chemical composition, and higher short- and long-range ordering. It is recognizable that the bottom-up approach is driven mostly by the reduction of Gibbs free energy (ΔG), so that the nanocrystalline materials are in a state closer to a thermodynamic equilibrium state. In contrast to this, top-down approach most likely insets internal stress, in addition to contaminations and surface defects. Attrition or ball milling is a generic top-down method in making nanostructures, whereas the colloidal dispersion or gas-based reduction is a usual example of bottom-up approach. The former produces polycrystalline structures with irreproducible crystallography and poorly controlled grain orientation. The latter has generated a reproducible collection of structures and architectures, including alloys, pure metals, anisotropic nanostructures, and core-shell. In lithography, the process may be assumed as a hybrid approach, since the thin film growth is bottom-up and etching is top-down, while nanolithography is generally a bottom-up approach.

Author details

Behrooz Movahedi
Department of Nanotechnology Engineering, Faculty of Advanced Sciences and Technologies, University of Isfahan, Isfahan, Iran

*Address all correspondence to: b.movahedi@ast.ui.ac.ir

IntechOpen

© 2019 The Author(s). Licensee IntechOpen. This chapter is distributed under the terms of the Creative Commons Attribution License (http://creativecommons.org/licenses/by/3.0), which permits unrestricted use, distribution, and reproduction in any medium, provided the original work is properly cited.

Section 2

Nanocrystalline Ceramics

Chapter 2

Silicon Nanocrystals and Amorphous Nanoclusters in SiO_x and SiN_x: Atomic, Electronic Structure, and Memristor Effects

Vladimir Volodin, Vladimir Gritsenko, Andrei Gismatulin and Albert Chin

Abstract

Semiconductor nanocrystals in dielectric films are interesting from fundamental aspect, because quantum-size effects in them appear even at room temperature, so such objects can be called as "quantum dots". Silicon nanocrystals and amorphous silicon nanoclusters in substoichiometric SiO_x and SiN_x films are traps for electrons and holes that apply in nonvolatile memory devices. In this chapter the formation of silicon nanocrystals and silicon amorphous nanoclusters in SiO_x and SiN_x films was studied using structural and optical methods. The phonon confinement model was refined to obtain sizes of silicon nanocrystals from analysis of Raman scattering data. Structural models that lead to nanoscale potential fluctuation in amorphous SiO_x and SiN_x are considered. A new structural model which is intermediate between random mixture and random bonding models is proposed. Memristor effects in SiO_x films are discussed.

Keywords: silicon suboxides, silicon subnitrides, nanocrystals, amorphous nanoclusters, phonon confinement model, nanoscale potential fluctuations, memristor

1. Introduction

Nanometer-sized semiconductor crystals, the so-called nanocrystals (NCs) and amorphous nanoclusters, embedded in wide-gap insulating matrices, have shown significant promises for application in nanoelectronics (nonvolatile memory) and optoelectronics (light-emitting diodes (LEDs)) [1]. Quantum effects in such hetero-systems are manifested even at room temperature. For example, a bright photoluminescence (PL) was observed in dodecyl-passivated colloidal Si NCs with external quantum efficiency (QE) up to 60% [2]. Since in some experiments single NCs originated delta-function-like energy photoluminescence spectra [3], they can be called as quantum dots. In metal-dielectric-semiconductor structures based on SiN_x films with Si nanoclusters, an effective electroluminescence with red, green, and blue light-emitting diodes was demonstrated [4]. Since NCs in a dielectric matrices act also as traps for charge carriers, the NCs' based structures have also

IntechOpen

perspectives to yield nonvolatile memory devices [5]. Last time the perspectives of application of nonstoichiometric SiO_x and SiN_x films with Si NCs and amorphous nanoclusters in memristors have arisen [6]. This chapter is devoted to formation, structural and optical studies of such films, and the development of new structural models which would explain the presence of nanoscale potential fluctuations in such nonstoichiometric films.

2. Forming of Si NCs and amorphous nanoclusters in dielectric films

There are several technological approaches for the fabrication of Si NCs and amorphous Si nanoclusters in various dielectric films: Si^+ ion implantation with consequent thermal annealing; co-sputtering of Si and SiO_2 targets on cool substrates; chemical vapor deposition (CVD) and plasma-enhanced chemical vapor deposition (PECVD) methods; evaporation of Si, SiO, or SiO_2 under high vacuum and their deposition onto cool substrates; evaporation of Si target in atmosphere with definite partial pressure of oxygen; and deposition on cool substrates. Each method has advantages and peculiarities.

The main advantage of ion implantation is a very precise control of the dose of the embedded silicon atoms—control of the projected range of silicon ions (it depends on the energy of ions). The disadvantage is the need for high-temperature annealing for the formation of silicon nanoclusters and even more high-temperature annealing (up to 1150°C) for their crystallization. So, this process cannot be "back-end-of-line" process in device production.

The benefits of different approaches that use co-sputtering are simplicity, the possibility to control stoichiometry using various intensity of evaporation of the targets, and the opportunity to use different substrates. The CVD and PECVD methods allow using large-scale substrates; the control of stoichiometry is possible using different ratios of reagent gases. The main advantage of PECVD is low temperature of the deposition process, but because almost all reagent gases contain hydrogen, the deposited SiO_x and SiN_x films are hydrogenated; in this case they should be marked as SiO_x:H and SiN_x:H films. Sometimes the hydrogen content is undesirable because it leads to instability of the characteristics of the films.

It should be noted that applying of various deposition methods leads to variation of the structural model of nonstoichiometric films. The structure of nonstoichiometric SiO_x or SiN_x films can be described in the framework of the random mixture (RM) or random bonding (RB) models [7]. In the RM model, SiO_x is treated as a mixture of two phases: the stoichiometric phase SiO_2 and the Si. In the RB model, SiO_x is assumed to consist of Si—O(ν)/Si(4 − ν) structural units, ν = 0, 1, 2, 3, or 4, in which Si atoms statistically substitute O atoms in each Si—O(4) structural unit. The structure of films formed by ion implantation and co-sputtering with ion beam evaporation of targets is closer to RB model; the structure of PECVD films can be closer to RM model. But structure of real films is always not pure RB or pure RM, and real structure of the films (which is the cause of nanoscale potential fluctuations in nonstoichiometric films) will be discussed below.

3. Raman scattering in SiO_x and SiN_x films: phonon confinement in Si NCs

Direct methods for studying the structure of nonstoichiometric films (such as high-resolution transmission electron microscopy (HRTEM)) are usually very time-consuming and destructive. Optical methods for studying the structure of NCs and amorphous nanoclusters are nondestructive and express. Among the optical methods, the most informative is inelastic light scattering—Raman scattering.

Raman scattering measurements are usually carry out to check the presence of a crystalline or amorphous Si phase in as-deposited and annealed SiO_x and SiN_x films. Due to the absence of long-range order and breaking of translation symmetry, the Raman spectrum of amorphous Si is an image of effective density of vibrational states for transversal optical (TO) and transversal acoustical (TA) modes and contains two broad peaks at approximately 480 and 150 cm^{-1}, correspondingly [8]. According to the quasi-momentum selection rules, in monocrystalline silicon, only phonons from the center of the Brillouin zone are active in Raman scattering; therefore the frequency of the Raman peak in this case is 520.5 cm^{-1} [9], and the full width on half maximum (FWHM) usually is about 5 cm^{-1}; it is much narrower than the width of amorphous peak. The intensity of the crystalline peak depends on the contents of the crystalline phase, and, using the analysis of experimentally measured integrated Raman scattering intensities I_c and I_a for c-Si and a-Si phases, one can obtain the volume part of crystalline phase in a two-phase film. The critical parameter of this method is the ratio of the integrated Raman cross section for c-Si to a-Si, $y = \Sigma_c/\Sigma_a$. Knowing this parameter, one can use the next equation:

$$\rho_c = I_c/\left(I_c + yI_a\right) \tag{1}$$

Recently, we have clarified the Bustarret data [10] on the dependence of the parameter y on the size of nano- and micro silicon crystals [11].

Due to softening of the conservation law of quasi-momentum in NCs, the short-wave phonons can take part in Raman scattering in this case. The confined in NCs phonons are characterized by a narrow peak at a position of 500–520 cm^{-1}. The position and the width of the peak strongly depend on the size and structure of the NCs [12, 13]. We have developed an improved phonon confinement model (PCM) for the analysis of average size of Si NCs from Raman data.

PCM allows us to calculate the Raman spectra for NCs of various sizes [12–14]. The physical entity of the model is the following. The eigenfunction of phonon with quasi-momentum $\hbar\mathbf{q}_0$ in an infinite crystal is

$$\Phi\left(\mathbf{q}_0, \mathbf{r}\right) = u(\mathbf{r})\ e^{i\,\mathbf{q}_0\,\mathbf{r}}, \tag{2}$$

where $u(\mathbf{r})$ has the periodicity of the lattice and $\mathbf{q}_0$ is the wavenumber of phonon.

The eigenfunction $\Psi\left(\mathbf{q}_0, \mathbf{r}\right)$ for a phonon confined in NCs is a function $\Phi\left(\mathbf{q}_0, \mathbf{r}\right)$ multiplied by the phonon weighting function $W(\mathbf{r},\ L)$ ("envelope" function for displacement of atoms in NCs); the weighting function depends on NC size L:

$$\Psi\left(\mathbf{q}_0, \mathbf{r}\right) = W(\mathbf{r}, L)\Phi\left(\mathbf{q}_0, \mathbf{r}\right) = \Psi'\left(\mathbf{q}_0, \mathbf{r}\right)u(\mathbf{r}), \tag{3}$$

where $\Psi'\left(\mathbf{q}_0, \mathbf{r}\right)$ is equal to $W(\mathbf{r},\ L)\ e^{i\,\mathbf{q}_0\,\mathbf{r}}$. The confined phonon can be described by a wave packet. To calculate the Raman spectrum, one should expand $\Psi'\left(\mathbf{q}_0, \mathbf{r}\right)$ in a Fourier presentation:

$$\Psi'\left(\mathbf{q}_0, \mathbf{r}\right) = \int C\left(\mathbf{q}_0, \mathbf{q}\right)\ e^{i\,\mathbf{q}\,\mathbf{r}}\ d^3\mathbf{q}. \tag{4}$$

The main task is the determination of Fourier coefficients $C\left(\mathbf{q}_0, \mathbf{q}\right)$ from an adequate physical model. The "weight" of the phonon with quasi-momentum $\hbar\mathbf{q}$ in the wave packet is proportional to $\left|C\left(\mathbf{q}_0, \mathbf{q}\right)\right|^2$.

The wavenumber of scattered photon is in our case about three orders of magnitude lower than the wavenumber of a phonon at the Brillouin zone boundary, so one can assume $\mathbf{q_0} \cong 0$. So, the first-order Raman spectrum is

$$I(\omega) \cong \int |C(0,\ \mathbf{q})|^2 \frac{n(\omega'(\mathbf{q})) + 1}{(\omega - \omega'(\mathbf{q}))^2 + (\Gamma/2)^2} d^3\mathbf{q} \tag{5}$$

where $n(\omega'(q)) + 1 = \frac{1}{e^{\hbar\omega'(q)/kT} - 1} + 1$ is the Bose-Einstein factor, $\omega'(\mathbf{q})$ is phonon dispersion, and Γ is FWHM of the Raman peak of a single phonon [12–14]. We also have taken into account that the vibration modes (phonons) with lower frequencies have higher amplitudes of vibration. Energy of vibration $\hbar\omega'$ is proportional to $\langle u^2 \rangle k$, where $\langle u^2 \rangle$ is standard deviation of an atom from equilibrium position k which is Hooke's coefficient for a bond, and $k = m\omega'^2$ (m is mass of an atom). So, one can derive $\langle u^2 \rangle \sim \hbar/\omega'$ (Eq. (2.26) in book [15]). This correction is substantial especially for phonons with large frequency dispersion, so we use the equation

$$I(\omega) \cong \int_0^1 |C(0, \mathbf{q})|^2 \frac{n(\omega'(\mathbf{q})) + 1}{\omega'(\mathbf{q}) \cdot \left((\omega - \omega'(\mathbf{q}))^2 + (\Gamma/2)^2\right)} d^3\mathbf{q} \tag{6}$$

It was shown [14] that using a Gaussian curve as eigenfunction for a confined phonon leads to more adequate results than experimental spectra.

Usually, it can be assumed that NCs have a spherical shape with diameter L. Therefore, in spherical coordinate system, the phonon weighting function $W(\mathbf{r},\ L)$ depends only on radius coordinate r and does not depend on angles. Assuming that at the boundary of NC ($r = L/2$) the phonon amplitude is equal to $1/e$ (the phonon amplitude at center of NC is equal to 1), one can obtain.

$$W(\mathbf{r}, L) = \exp\left(-4r^2/L^2\right), \text{ so } C(0, \mathbf{q}) \cong \exp\left(-\frac{|\mathbf{q}|^2 L^2}{16}\right) \ |C(0, q)|^2 \cong \exp\left(-\frac{q^2 L^2}{8}\right) \tag{7}$$

Usually, only empirical expressions for phonon dispersion were used in PCM [12, 13]. But the empirical expressions are accurate enough only near the Brillouin zone center. Also, in earlier approaches, the differences between dispersions of longitudinal optic (LO) and transverse optic (TO) phonons were usually not taken into account. In general, for crystals with diamond-type lattice, there are six phonon branches with dispersions $\omega_i'(q)$, so the first-order Raman spectrum for phonon weighting function $W(\mathbf{r},\ L) = \exp\left(-4r^2/L^2\right)$ is

$$I(\omega) \cong \sum_{i=1}^{6} \int_0^{q\max} \exp\left(-\frac{q^2 L^2}{8}\right) \frac{n(\omega_i'(q)) + 1}{\omega_i'(\mathrm{q}) \cdot \left((\omega - \omega_i'(\mathrm{q}))^2 + (\Gamma/2)^2\right)} q^2 dq. \tag{8}$$

Wavenumbers are varied from 0 up to q_{max} (edge of the Brillouin zone). For directions with high symmetry (<100> and <111>), it should be noted that some phonon branches are degenerated. The density of states for phonons is proportional to $q^2 dq$.

In some approaches, the phonon frequencies are determined using "ab initio" quantum mechanical calculations [16], but this method requires large computational resources, while NCs with diameters >3 nm contain more than 1 thousand atoms. So, for calculation of phonon dispersion, the Keating model of valence forces [17] was used. In this simple but adequate model, the elastic energy of the crystal depends on bond length and on deviation of bond angle from ideal tetrahedral angles. We consider atom-atom interaction only between the nearest neighbor. For a crystal with diamond-type lattice, the elastic energy of unit cell is

$$E = \frac{3}{16}\sum_i \sum_j \frac{k_l}{a^2}\left(\left(\vec{r}_i - \vec{r}_j\right)^2 - \frac{3a^2}{16}\right)^2 + \frac{3}{8}\sum_i \sum_{k,j>k}^{4} \frac{k_\varphi}{a^2}\left(\left(\vec{r}_i - \vec{r}_j\right)\cdot\left(\vec{r}_i - \vec{r}_k\right) + \frac{a^2}{15}\right)^2\Bigg], \quad (9)$$

where k_l and k_ϕ are elastic constants (Hooke's coefficients) and a is lattice constant. TO and LO phonons at the Brillouin zone center are degenerated for crystals with diamond-type lattice. The frequency is given by

$$\omega_\Gamma = \sqrt{\frac{8\left(k_l + 3k_\varphi\right)}{3m}} \quad (10)$$

where m is the mass of Ge atoms. As it was mentioned above, Si frequency of TO and LO phonons at the Brillouin zone center is equal to 520.5 cm^{-1} So, the elastic constants k_ϕ and k_l are not independent [see Eq. (10)]. The elastic constant k_l was determined from approximation of calculated dispersions in directions <100>, <110>, and <111>, obtained from neutron scattering data [18, 19]. It is important to consider phonons of different directions, because in experiment, Raman signal comes from a large amount of randomly oriented NCs, and all phonon modes are intermixed. The exact expressions for phonon dispersions in directions <100>, <110>, and <111> for Keating model are published in Ref. [14] and are very cumbersome. To calculate the first-order Raman spectrum, one should use these dispersions in the Eq. (8). Dispersion in different directions should be used with its corresponding weight. There are 6 physically equivalent <100> directions, so the weight of this dispersion is 6. Similarly the weight of dispersion along <111> and <110> directions are, respectively, equal to 8 and 12. Thus, all calculations were performed with the phonon dispersion in the Keating model, taking into account the phonon dispersion for the three main directions in Si.

Figure 1 shows the results of calculations of the Raman spectra of Si NCs of different diameters using improved PCM. It is seen that for Si NC with diameter of 10 nm, the effect of phonon confinement is significant. The peak shifted and broadened relative to the peak from the bulk Si. For sizes below 10 nm, the NCs' Raman spectrum becomes asymmetric.

Figures 2 and **3** summarize the results of calculations compared with experimental results. **Figure 2** shows the difference between the position of Raman peaks of Si NCs and bulk Si. The average sizes of the Si NCs were determined from HRTEM data. As can be seen, the results of calculations in improved PCM agree well with experimental data but have some differences from the simulation results presented in earlier works [12, 13, 16]. It should be noted that results of calculations in the improved PCM are adequate for a broad range of Si NCs' sizes (from 3 to 10 nm). Note, however, that if during measurements, the heating of the sample under laser spot takes place, the Raman peak will shift (due to anharmonicity of phonons). If the system contains mechanical stress, it will also cause a shift of the Raman peak [14]. **Figure 3** shows the dependence of the Raman peak width with the size of the NCs, for calculation with improved PCM and for experimental data. Some differences between the experimental data and calculations are visible. In particular, large width of the experimental spectra, compared with the calculated spectra, may be due to the dispersion of the size of the Si NCs. Thus, if anharmonicity effect due to heating or mechanical stress is not relevant, the present

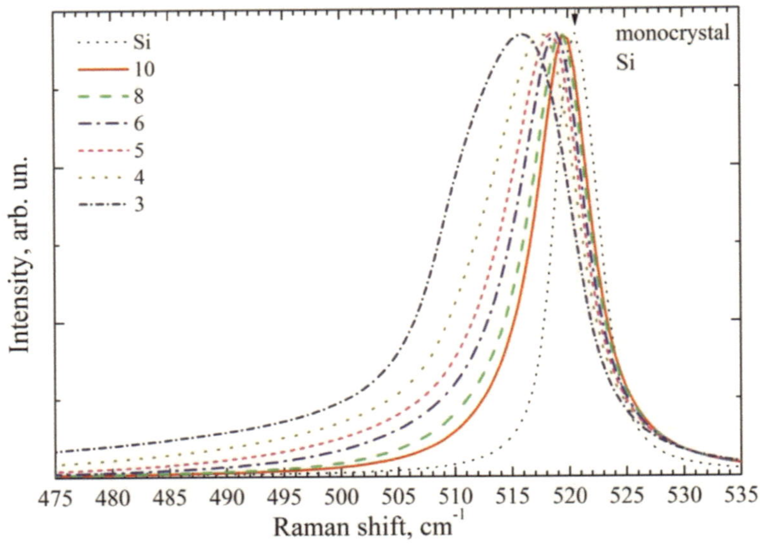

Figure 1.
Calculated Raman spectra of Si NCs of diameters from 10 to 3 nm.

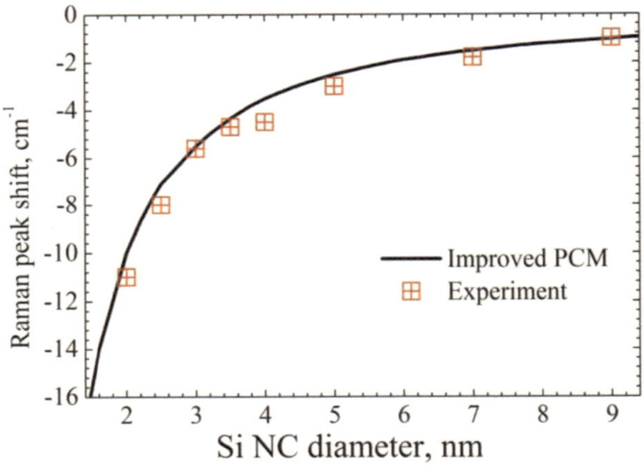

Figure 2.
Shift of the position of the Raman peak for optical phonons confined in Si NCs of various sizes is shown relative to position of Raman peak for bulk Si. The solid curve represents the results obtained using improved PCM (dispersion is calculated in the Keating model taking into account the angular phonon dispersion); red crosses show the experimental data.

improved model allows us to determine the average size of the Si NCs from the analysis of the Raman spectra for a wide range of sizes.

Experimental data on **Figures 2** and **3** were obtained for Si NCs in SiO_x, SiN_x, and amorphous Si matrix and also for free-standing Si nanopowders. This indicates that the PCM is adequate for various matrices; the main demand is that localized phonons in NCs strongly damp in matrix.

In **Figure 4** the Raman spectra of PECVD-deposited SiO_x:H films are shown. Deposition was made from the mixture of monosilane (SiH_4) diluted by argon (Ar) and oxygen (O_2) diluted by helium (He). The stoichiometry parameter “x” was changed by varying of oxygen concentration. The temperature of Si (100) monocrystalline substrate was 200°C. The thickness of SiO_x:H films was about 200 nm. The value of stoichiometry parameter “x” was obtained from the analysis of X-ray photoelectron spectroscopy (XPS) data.

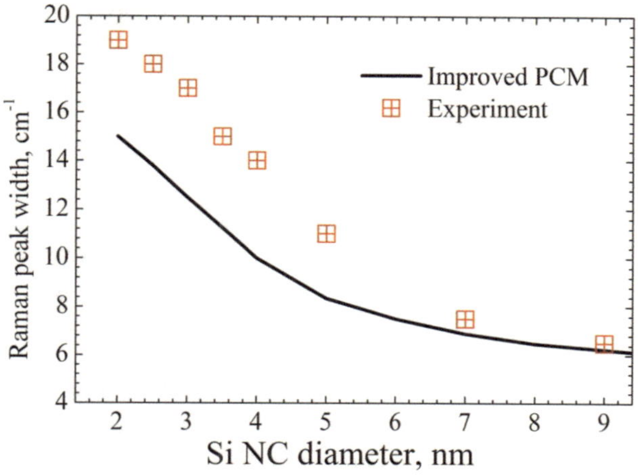

Figure 3.
Width of the Raman peak for optical phonons confined in Si NCs of various sizes. The solid curve represents the results of calculation; red crosses show the experimental data.

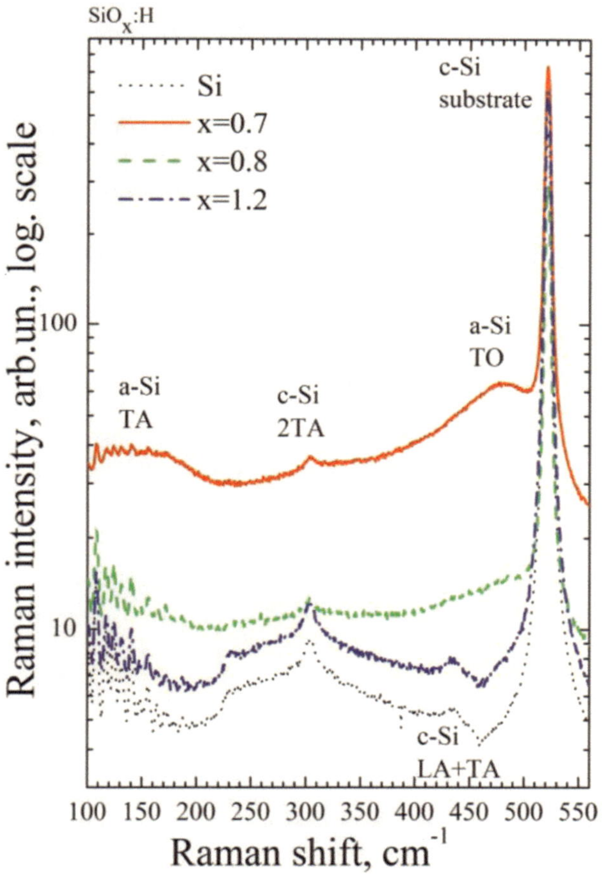

Figure 4.
Raman spectra of SiO_x:H films of different stoichiometry.

Raman spectra were registered at room temperature in back-scattering geometry. For excitation, the 514.5-nm line of an Ar^+ laser was used. No polarization analysis for scattered light was performed. A Horiba Jobin Yvon T64000 spectrometer was used for measuring Raman spectra with a spectral resolution better than 2 cm^{-1}. A special facility for microscopic Raman studies was also employed. The laser-beam power reaching the sample was 2 mW. For minimization of the heating of the structures under the laser beam, the sample was placed somewhat below the focus in a situation in which the laser-spot size was equal to 10 μm.

So, registered Raman spectra of the SiO_x:H films and the Raman spectrum of a monocrystalline silicon substrate are shown in **Figure 4**. Evidently, a very intense signal due to silicon substrate is observed; this is a line due to the 520.5 cm^{-1} long-wave optical phonon. For clarity, the vertical scale is plotted logarithmic. Besides, features originating from two-phonon scattering phenomena, namely, those due to events involving two acoustic phonons (2TA ~300 cm^{-1}, LA + TA ~425 cm^{-1}), were observed in the spectrum of single-crystal silicon. SiO_x:H films are semitransparent ones in the visible light, and their spectrum also exhibits a signal due to the substrate. The narrow peaks with wavenumbers lower than 160 cm^{-1} resulted from the inelastic scattering of light by atmospheric molecules. As it was mentioned, in Raman spectrum of amorphous Si clusters, there are TO (480 cm^{-1})- and TA (150 cm^{-1})-related broad peaks.

In **Figure 4** one can see that the SiO_x:H films with $x < 1$ contain noticeable amount of amorphous Si clusters. In the spectrum of film with $x = 1.2$, the TO and TA peaks are practically absent. If the structure of this film corresponded to the RM model, then a significant number of clusters of amorphous silicon would be present in it.

A similar picture is observed for PECVD-grown SiN_x:H films (**Figure 5a**).

The studied SiN_x:H films of different stoichiometric composition were grown using PECVD from a mixture of ammonia (NH_3) and monosilane (SiH_4) on Si substrates with orientation (001). It is known that the composition of SiN_x films ($0 < x < 4/3$) depends on the NH_3/SiH_4 flow ratio. The temperature of the substrates during deposition was 150°C. The value of stoichiometry parameter "x" was defined using of XPS data.

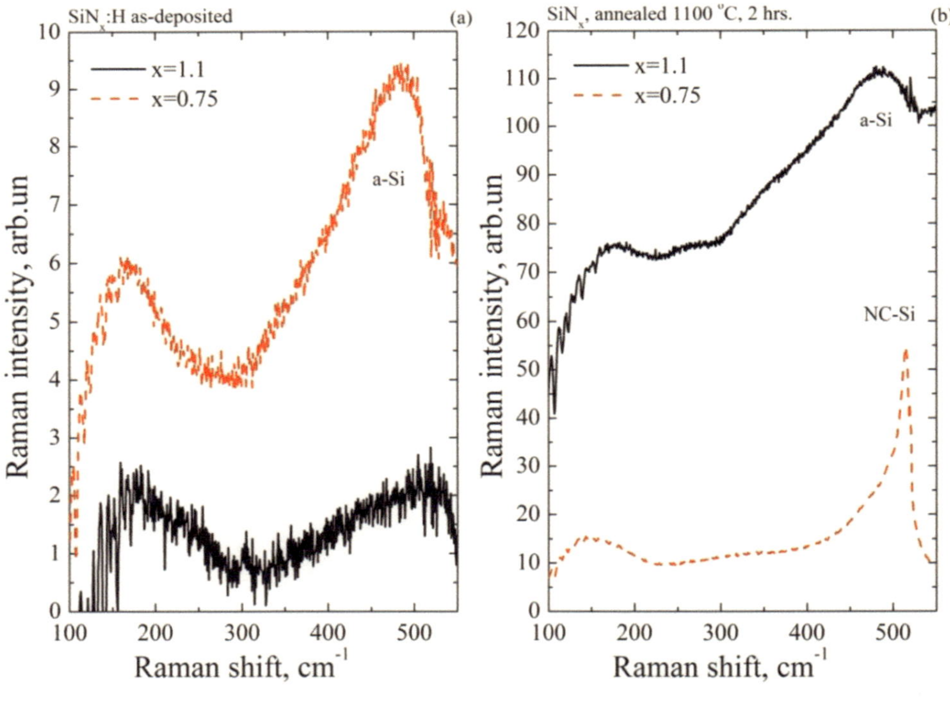

Figure 5.
Raman spectra of SiN_x:H films of different stoichiometry: (a) as-deposited films and (b) as-annealed films.

One can see in Raman spectra of as-deposited SiN_x:H films the amorphous Si peaks (**Figure 5a**). It should be noted that the spectrum of Si substrate was subtracted from spectra of studied structures. One can see that the SiN_x:H films with x < 1 contain noticeable amount of amorphous Si clusters. In the as-deposited sample with x = 1.1, the signal from amorphous Si is present, but small

In **Figure 5b** the spectra of SiN_x films after annealing at Ar atmosphere (1100°C, 2 hours) are presented. One can see that in spectrum of sample with low concentration of Si (x = 1.1), the annealing leads to growth of TO and TA peaks related to amorphous Si. It means that the annealing contributed to the gathering of excess silicon atoms into amorphous clusters and the structure of annealed film is close to RM model. Nevertheless, even such a high-temperature annealing did not lead to crystallization of amorphous nanoclusters. In spectrum of $SiN_{0.75}$ film, there is narrower peak with position 514.5 cm^{-1}. The shift compared with position of peak of monocrystalline Si is about 6 cm^{-1}. According to the data presented in **Figure 2**, such shift is corresponding to Raman scattering by optical phonons localized in Si NCs with average size about 3 nm. It is worth to note that this size is closed to critical size of stable crystalline nuclei of Si.

The photoluminescence under excitation with ultraviolet laser HeCd laser (λ = 325 nm) was also studied in as-deposited and annealed SiN_x:H films. Annealing leads to an increase in the intensity of the photoluminescence, apparently due to the annealing of non-radiative defects. The maximum of the photoluminescence signal shifted to the long-wavelength direction (redshift) with an increase in the content of excess silicon in silicon nitride films.

4. IR absorption in SiO_x and SiN_x films: the evidence of deviation from the RM model

The SiO_x:H and SiN_x:H films were studied using Fourier transform infrared (FTIR) absorption spectroscopy; the spectrometer FT-801 having a spectral resolution of 4 cm^{-1} was used.

The IR spectra of SiO_x samples in **Figure 6** show an absorption peak on the stretching vibrations of the Si—O bonds (TO_3 peak [20]). Pai et al. [21] found that the position of this peak (in inverse centimeters) in SiO_x films almost linearly depends on the stoichiometry parameter x, like.

$$\nu = 925 + 75x \tag{11}$$

From the data of **Figure 6**, it can be seen that the position of the TO_3 peak for the studied samples varies from 1040 to 1060 cm^{-1}. So, according to Eq. (11), the expected stoichiometry of silicon oxide SiO_x should change only slightly. But, according to XPS data, the stoichiometry of the SiO_x films varies widely (from 0.7 to 1.2). This suggests that the structure of our films does not correspond to the RB model (otherwise, the shift range of the TO_3 peak position would be much wider). However, the structure of our films does not correspond to the RM model either (the position of the TO_3 peak for all the films would correspond to the SiO_2 matrix and would be about 1075 cm^{-1}). It is worth also to note that the peaks corresponding to absorption by Si—H and O—H bonds were observed in the films, so the as-deposited SiO_x films are hydrogenated.

Figure 7a shows the IR absorption spectra of as-deposited SiN_x:H films as well as silicon substrate. Nonpolar Si—Si bonds that are active in the Raman process are not active in the absorption process, but Si—N, Si—H, and N—H bonds are active in it, which makes it possible to obtain information on the structure of a-SiN_x:H films

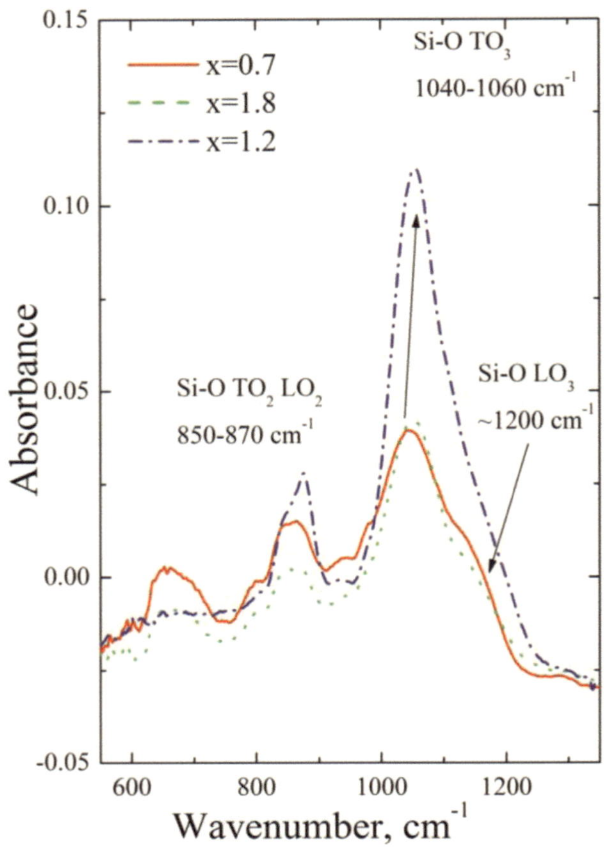

Figure 6.
IR-absorbance spectra of SiO_x:H films of different stoichiometry.

using the IR absorption method. Optical density A (natural logarithm of 1/T, where T is transmission) is plotted on the vertical axis. In the spectra of the films grown at high ratio of ammonia to monosilane fluxes (x = 1.3), absorption peaks at 3340 cm^{-1} are visible. This is the absorption on the stretching vibrations of the nitrogen-hydrogen bonds [22]. In the spectra of the films grown at a ratio of ammonia to monosilane fluxes of 1 and 0.5 (x = 1.1 and 0.75 accordingly), the intensity of this peak is very low, which means that the concentration of hydrogen bound to nitrogen decreases with increasing concentration of excess silicon. It is also seen from **Figure 7** that the spectra of samples containing excess silicon contain a peak with a position of 2150 cm^{-1}. This is the peak from absorption on the stretching vibrations of the silicon-hydrogen bonds [22]. The intensity of this peak depends on the ammonia/monosilane ratio; the intensity is very low in Si_3N_4 film, but it grows with increasing concentration of excess silicon. The peak at ~1100 cm^{-1} observed in all spectra corresponds to the absorption of stretching vibrations of silicon-oxygen bonds in the silicon substrate. These bonds also give peaks from 400 to 800 cm^{-1} of twisting, wagging, rocking, and scissor modes. In addition, the spectrum of monocrystalline silicon contains peaks from multiphonon lattice absorption in silicon itself (features in the region of 607–614 cm^{-1}). In some spectra, there is also a "parasitic" peak with a position of 2350–2400 cm^{-1} associated with absorption on carbon dioxide gas (in the process of measuring its concentration slightly changed, as a result it was not completely

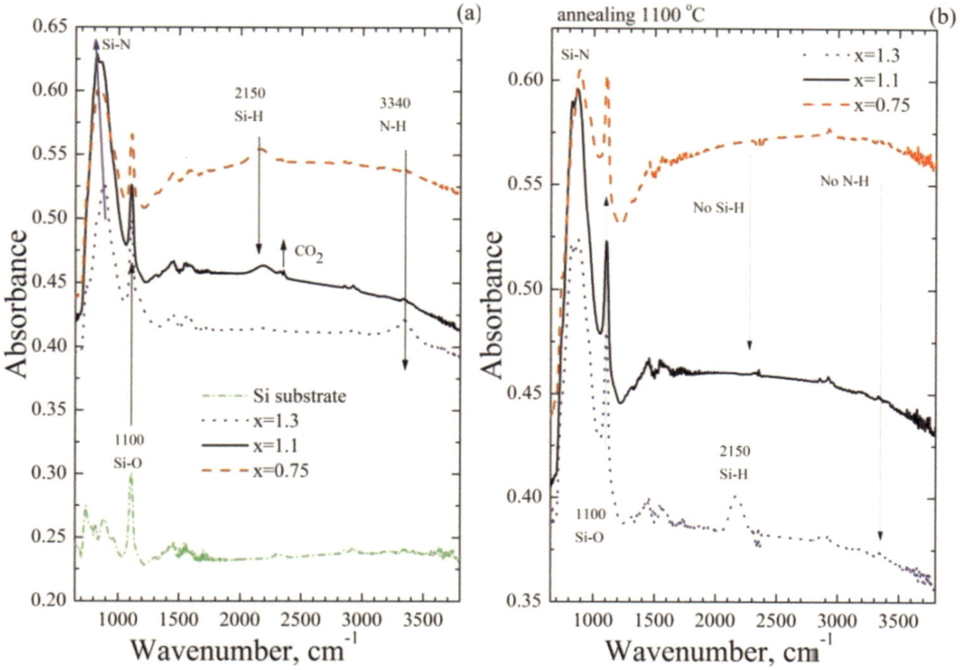

Figure 7.
IR-absorbance spectra of SiN_x:H films of different stoichiometry: (a) as-deposited films and (b) as-annealed films.

removed when dividing the spectrum from the samples by the reference spectrum of the air).

Let us turn to the absorption peak due to vibrations of the Si—N bonds in **Figure 7**. In the spectra of all the films, there are peaks from the stretching vibrations of these bonds. The spectra were approximated by Gaussian curves, and peak positions were determined. The position of the absorption peaks on the stretching vibrations of the Si—N bonds is shifted, depending on the stoichiometry, from 880 to 860 cm^{-1}. This effect was previously known (see work [23] and references therein). The general dependence is that the oscillation frequency decreases with decreasing stoichiometric parameter x in a-SiN_x film. This is observed in our experiment and again confirms that the structure of the films cannot be considered only within the framework of the RM model (in which the stoichiometry of the matrix surrounding the silicon inclusions is unchanged—the matrix parameter x is 4/3).

Figure 7b shows the IR absorption spectra of annealed (Ar atmosphere, 1100°C, 2 hours) SiN_x:H films. One can see that annealing leads to evaporation of hydrogen, except nearly stoichiometry (x = 1.3) film. In that film the Si—H peak becomes even more intensive after annealing. So, hydrogen has been removed from N—H bonds to Si—H bonds. This effect has already been observed in the work [24]. In the work [24], it was shown that in order to remove hydrogen from Si—H bonds, it is necessary to apply high-temperature annealing at very high pressure. It should be noted that in annealed films the position of the absorption peaks on the stretching vibrations of the Si—N bonds is also shifted, depending on the stoichiometry. The lesser parameter x, the higher is frequency of stretching vibrations of the Si—N bonds.

So, the analysis of IR absorption data is an evidence of deviation of structure of real SiO_x and SiN_x films from the RM model.

5. New structural model explaining nanofluctuations of potential in SiO_x and SiN_x

Nonstoichiometric silicon oxide SiO_x and nitride SiN_x are tetrahedral compounds whose structure is defined by the Mott octahedral rule [7, 25]. SiO_x and SiN_x are synthesized under thermodynamically nonequilibrium conditions. Therefore, the structure of nonstoichiometric SiO_x and SiN_x layers depends on synthesis conditions, i.e., temperatures, gas pressure, and annealing.

The RB and RM models [7] are two extreme cases of the SiO_x and SiN_x structure description. As a rule, at low synthesis temperatures (<300°C), its structure is described by the RB model; higher synthesis temperatures promote phase separation in SiO_x layers, i.e., such layers should be better described by the RM model.

According to the RB model [26], the probability of finding the $SiO_vSi_{(4-v)}$ tetrahedron (the fraction of given v-type tetrahedra), where v = 0, 1, 2, 3, 4 in SiO_x for composition x, is given by.

$$W_{\nu}^{RB}(x) = \frac{4!}{\nu!(4-\nu)!}\left(\frac{x}{2}\right)^{\nu}\left(1-\frac{x}{2}\right)^{4-\nu} \tag{12}$$

According to the RM model [7], the calculated spectrum consists of two tetrahedron types, SiO_4 and $SiSi_4$. The fraction of SiO_4 and $SiSi_4$ tetrahedra in the calculation of Si 2p spectra is (1 − x/2) and x/2, respectively. To simulate the photoelectron spectrum *I(E)*, the W_v peaks obtained using the RB and RM models were broadened by the Gaussian using the formula:

$$I(E) = \sum_{\nu} W_{\nu} e^{-(E-E_{\nu})^2/2\sigma_{\nu}^2} \tag{13}$$

where E_v and σ_v are the peak energy and "half-width" for a given tetrahedron type. The SiO_x film composition was calculated assuming that the calculated spectrum is a superposition of five peaks corresponding to five $SiO_vSi_{(4-v)}$ tetrahedra, v = 0, 1, 2, 3, 4. The fraction of tetrahedral was selected from the best fit of the spectrum *I(E)* calculated by Formula (12).

Figure 8a shows five experimental photoelectron spectra of the Si 2p level in SiO_x. We can see that the energy and half-width of the Si_{4+} peak belonging to the a-SiO_2 phase and the Si peak belonging to the a-Si phase are E_0 = 103.5 eV and σ_0 = 1.2 eV and E_4 = 99.5 eV and σ_4 = 0.6 eV, respectively. The position and half-width for Si_{3+}, Si_{2+}, and Si_+ peaks ($SiSiO_3$, $SiSi_2O_2$, and $SiSi_3O$ tetrahedra) were determined using linear interpolation of E_0, E_4, σ_0, and σ_4 using the number of oxygen atoms as a parameter. Dashed curves in **Figure 8a** show the spectra calculated from the best fit with experimental spectra. The calculation for the SiO_x film composition predicts the following values, x = 0, 0.7, 0.98, 1.47, and 2.0.

Figure 8b shows the experimental photoelectron spectra of the Si 2p level in SiO_x and the results of RB model simulation. The RB model predicts a single peak being a superposition of five peaks corresponding to five $SiO_vSi_{(4-v)}$ tetrahedra (v = 0, 1, 2, 3, 4). The calculated peak shifts to lower binding energies with decreasing oxygen concentrations. The position and half-width of the calculated SiO_x peak for x = 0.7, 0.98, 1.47 is not in agreement with the experimental spectrum

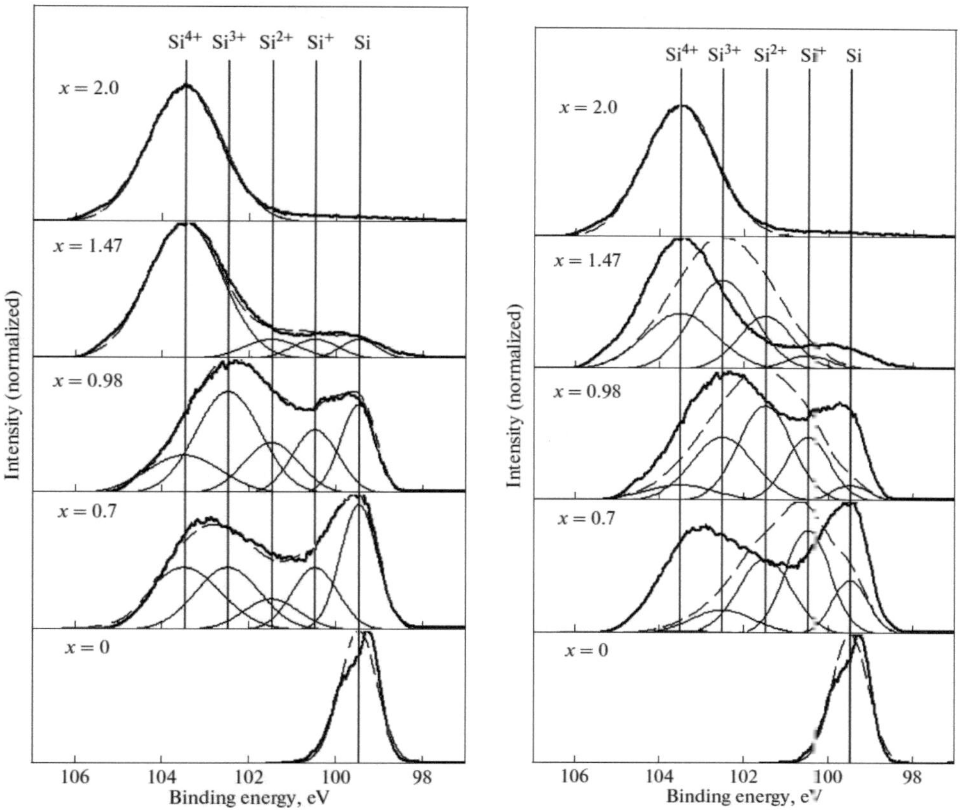

Figure 8.
(a) (left) Experimental photoelectron spectra of the Si 2p level in SiO_x (bold curve) and simulation results for determining the film composition (dashed curve). Symbols Si_{4+}, Si_{3+}, Si_{2+}, Si, and Si indicate intensities of the Si 2p level for SiO_4, $SiSiO_3$, $SiSi_2O_2$, $SiSi_3O$, and $SiSi_4$ tetrahedra, respectively. (b) (right) Experimental photoelectron spectra of the Si 2p level in SiO_x (bold curve) and the results of simulation using the RB model (dashed curve). Symbols Si_{4+}, Si_{3+}, Si_{2+}, Si_{+}, and Si indicate intensities of the Si 2p level for SiO_4, $SiSiO_3$, $SiSi_2O_2$, $SiSi_3O$, and $SiSi_4$ tetrahedra, respectively.

of the Si 2p level. The RB model underestimates the role of $SiSi_4$ and SiO_4 tetrahedra in calculating the intermediate SiO_x composition (x = 0.7, 0.98, 1.47). Thus, there are five $SiO_vSi_{(4-v)}$ tetrahedron types in SiO_x; however, the probability of their detection is not quantitatively described by the RB model.

The calculation of experimental spectra using the RM model is shown in **Figure 9a**. According to the RM model, the calculation predicts the existence of two peaks corresponding to SiO_4 and $SiSi_4$ tetrahedra. The calculated energies of the Si 2p level peaks correlate with those of experimental spectra. The calculated spectra underestimate the contribution of $SiSiO_3$, $SiSi_2O_2$, and $SiSi_3O$ tetrahedra which exist in experimental spectra. For example, for $SiO_{0.98}$, the calculation predicts the presence of the SiO_2 phase which is not observed in experimental spectra. Thus, the RM model also does not describe the experimental photoelectron spectra.

We note that it is impossible to describe experimental spectra by summing the RB and RM spectra in corresponding proportions. This is easily seen in the case of the composition x = 0.7 for which both RB and RM models predict a significantly smaller content of the SiO_2 phase than it is observed in the experiment.

Figure 9b shows the photoelectron spectra of the valence band of SiO_x of variable composition, measured at an excitation energy of 1486.6 eV. At such an excitation energy, silicon states make the main contribution to the valence band

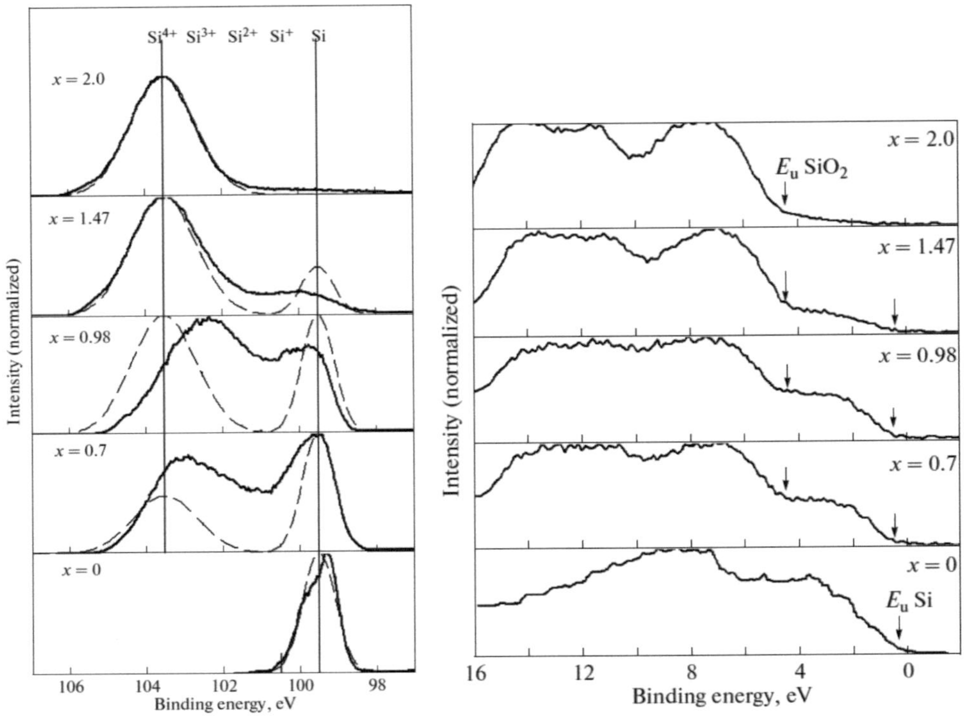

Figure 9.
(a) (left) Experimental photoelectron spectra of the Si 2p level in SiO_x (bold curve) and the results of simulation using the RM model (dashed curve). Symbols Si_{4+} and Si indicate intensities of the Si 2p level for SiO_4 and $SiSi_4$ tetrahedra, respectively. (b) (right) Experimental XPS data of the valence band of SiO_2, SiO_x, and Si. The top of the silicon valence band is taken as the reference point.

spectrum. Oxygen states (O 2p) at such excitation energies make a small contribution due to a low photoionization cross section. The SiO_x ($x > 0$) photoelectron spectrum contains three distinct peaks. As the silicon contents in SiO_x increase above the top of the silicon valence band (E_v Si), states caused by silicon appear (**Figure 9b**). The low-energy peak at 0–4 eV is caused by Si 3p orbitals in amorphous silicon. The peaks at energies above 4 eV are caused by Si 3s and 3p orbitals. These results independently point to the fact that SiO_x contains SiO_2 and Si.

To describe the SiO_x structure, it was proposed to use the intermediate model (IM). The IM model assumes local fluctuations of the SiO_x chemical composition, which result in bandgap fluctuations. For example, in [27], it was shown that the chemical composition of silicon oxide films can be identical, $SiO_{1.94}$, while the bandgap can vary in the range of 5.0–7.5 eV. **Figure 10a** shows the SiO_x energy-level diagram for section *A–A*. The horizontal line ($E = 0$) is electron energy reference point (the energy of vacuum). Symbols E_c and E_v denote the bottom of the conduction band and the top of the valence band in SiO_x. The SiO_2 bandgap is 8 eV [7]. Bandgap narrowing indicates a local increase in the silicon concentration in SiO_x. The least bandgap ($E_g = 1.5$ eV) corresponds to the silicon phase. Thus, the maximum scale of potential fluctuations for electrons and holes is 2.6 and 3.8 eV, respectively. **Figure 10a** shows all possible versions of the SiO_x spatial structure. White, black, and gray colors correspond to SiO_2, a-Si, and silicon suboxides, respectively. If the silicon cluster size *L* is small, size quantization effects can be observed in it. Such a cluster is indicated in the figure by numeral *1*. Numeral *2* indicates the macroscopic silicon cluster in silicon oxide. In this case, the intermediate layer of silicon suboxides is absent, and the Si/SiO_2 interface in the

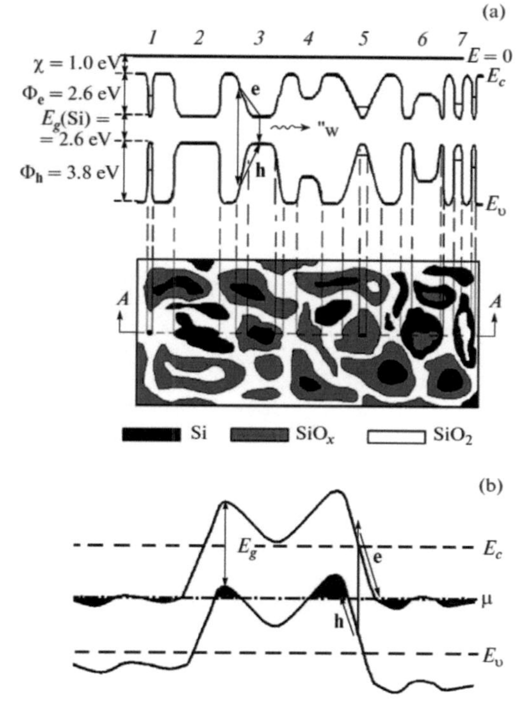

Figure 10.
(a) IM model: the schematic two-dimensional pattern of the SiO_x structure and the SiO_x energy-level diagram for section A–A. White, black, and gray colors correspond to SiO_2, a-Si, and silicon suboxides, respectively. Φ_e and Φ_h barriers for electrons and holes at the interface a-Si/SiO_2, respectively. (b) Model of potential fluctuation (Shklovskii-Efros model) in a heavily doped compensated semiconductor; μ is the Fermi level.

energy-level diagram boundary is sharp. Numeral *3* indicates the silicon cluster surrounded by silicon suboxide. In this case, the Si/SiO_2 interface in the energy-level diagram is shown by a smooth curve. Hereafter, it is assumed that the intermediate region (silicon suboxide) size significantly exceeds the Si—O and Si—Si bond length. Numeral *4* indicated the suboxide silicon cluster in SiO_2. Numeral *5* indicates the silicon cluster in silicon suboxide. Numeral *6* indicates the suboxide cluster in silicon, and numeral *7* indicates the SiO_2 cluster in silicon.

The model of large-scale potential fluctuations (Shklovskii-Efros model) in a heavily doped compensated semiconductor was developed earlier (**Figure 10b**) [28]. In this model, the bandgap is constant, and potential fluctuations occur due to the nonuniform spatial distribution of the charged ionized donors and acceptors. An electron-hole pair excitation results in spatial separation of electrons and holes, which does not facilitate their recombination. The principal difference between the proposed model of nanoscale potential fluctuations in SiO_x and the Shklovskii-Efros model is as follows. Large-scale potential fluctuations in compensated semiconductors are of electrostatic nature associated with spatial fluctuations of the charge density of donors and acceptors. The bandgap is constant (**Figure 10b**), and the electric field caused by spatial potential fluctuations promotes electron and hole separation. In SiO_x, potential fluctuations are caused by local fluctuations of the chemical composition.

In the IM model, in contrast to the Shklovskii-Efros model, the space charge is absent. According to the IM model, local fluctuations in the SiO_x chemical composition result in spatial potential fluctuations which, in turn, lead to changes in local electric fields for electrons and holes. These fields at the same point of the SiO_x sample are different in magnitude and direction (**Figure 10a**). When an

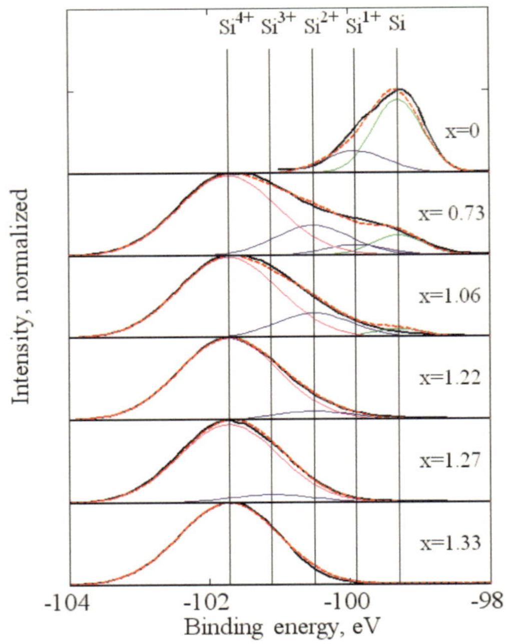

Figure 11.
Experimental XPS spectra of the Si 2p level in SiN_x (solid black lines) and the results of theoretical modeling using the IM model (dashed red lines). Green line is peak from Si—Si_4 tetrahedron, and magenta line is peak from Si—N_4 tetrahedron.

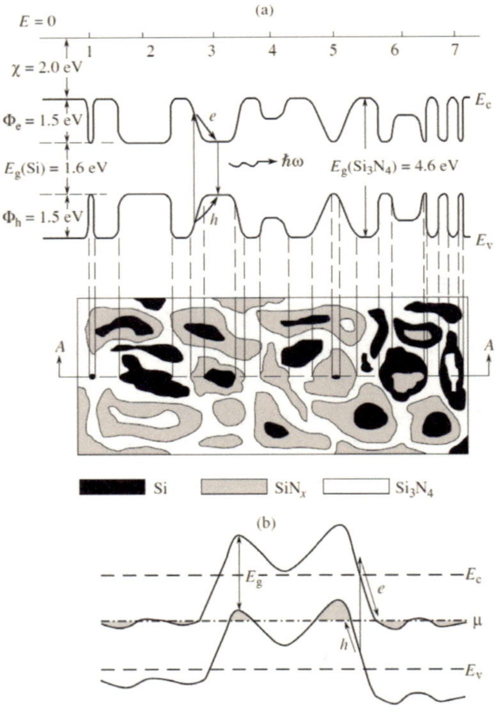

Figure 12.
Schematic diagrams illustrating the proposed intermediate model of SiN_x: (a) a two-dimensional diagram of SiN_x structure showing (bottom) the regions of a silicon phase, stoichiometric silicon nitride, and subnitrides and (top) the energy band profile of SiN_x in the A–A section (E_c is the conduction band bottom; E_v is the valence band top; Φ_e and Φ_h are the energy barriers for electrons and holes at the a-Si–Si_3N_4 interfaces, respectively; E_g is the bandgap width). (b) The potential fluctuations in Shklovskii–Efros model.

electron-hole pair is excited in SiO_x, the electric field for electron and hole promotes (see in **Figure 10a**) their recombination. In the case of the radiative recombination mechanism, SiO_x is an efficient emitting medium. Nanoscale potential fluctuations in SiO_2 promote electron and hole localization in potential wells (silicon clusters) [7]. This effect is used for developing the high-speed nonvolatile memory based on charge localization in SiO_x and can be used in memristors.

In the case of SiN_x films, the approach for determining the stoichiometric parameter x from XPS data analysis is similar, but unlike Eq. (12), the probability of finding the $SiN_vSi_{(4-v)}$ tetrahedron (the fraction of given v-type tetrahedra), where v = 0, 1, 2, 3, 4 in SiN_x for composition x, is given by

$$W_v^{RB}(v,x) = \frac{4!}{v!(4-v)!}\left(\frac{3x}{4}\right)^v\left(1-\frac{3x}{4}\right)^{4-v} \tag{14}$$

Experimental XPS spectra are also not described by pure RB or RM models. However, good agreement between the experimental and calculated spectra is observed for IM model (**Figure 11**).

The nanoscale fluctuations of potential in SiN_x films (**Figure 12**) are also similar to nanoscale fluctuations of potential in SiO_x. In the case of SiN_x films, the IM is more adequate to describe the real structure and fluctuation of potential. In schematic picture in the bottom of **Figure 12**, one can see the possible appearance of such structures—Si core surrounded by SiN_x shell and Si_3N_4 matrix. So, the proposed IM also can be called as core-shell-matrix model.

6. Memristor effects in SiO_x films

Resistive random-access memory (RRAM) [29, 30] is the highly promising candidate for the next-generation nonvolatile memory (NVM), because conventional charge-based memories, namely, dynamic random-access memory and flash memory, have too low capacitance after continuously downscaling into 1X-nm regimes. In addition, an RRAM array can be fabricated in the back end of the line of a complementary metal-oxide-semiconductor circuit, which makes such device an excellent candidate for embedded NVM (e-NVM) application. The typical write speed of RRAM device ranges from 100 ns to 1 μs, which is three to four orders of magnitude faster than flash memory. Such high-speed and process-compatible e-NVM can enable hardware technologies such as artificial intelligence and neuromorphic computing.

The conduction mechanism of RRAM, however, is not fully understood, and it is generally attributed to metallic filament conduction because of its metal-insulator-metal (MIM) structure, where the insulator is usually formed by metal oxide-based dielectric. The first RRAM that does not contain any metal in both the electrodes and dielectric insulator (nonmetal RRAM) is demonstrated here. To obtain RRAM device, a 15-nm-thick SiO_x was deposited directly on a p^+-Si substrate by reactive sputtering. Then, a 15-nm-thick amorphous n^+-Si layer was formed as the top junction electrode. The value x in SiO_x was determined to be 0.62. Because no metal or metallic ions were present in the whole RRAM device, metallic filaments were not formed.

Figure 13(a) depicts the measured *I-V* characteristics of an n^+-Si/$SiO_{0.62}$/p^+-Si RRAM device. During the forming step, the device was first subjected to a 6 V and 100 μA compliance current stress to attain the LRS. The same device was reset into

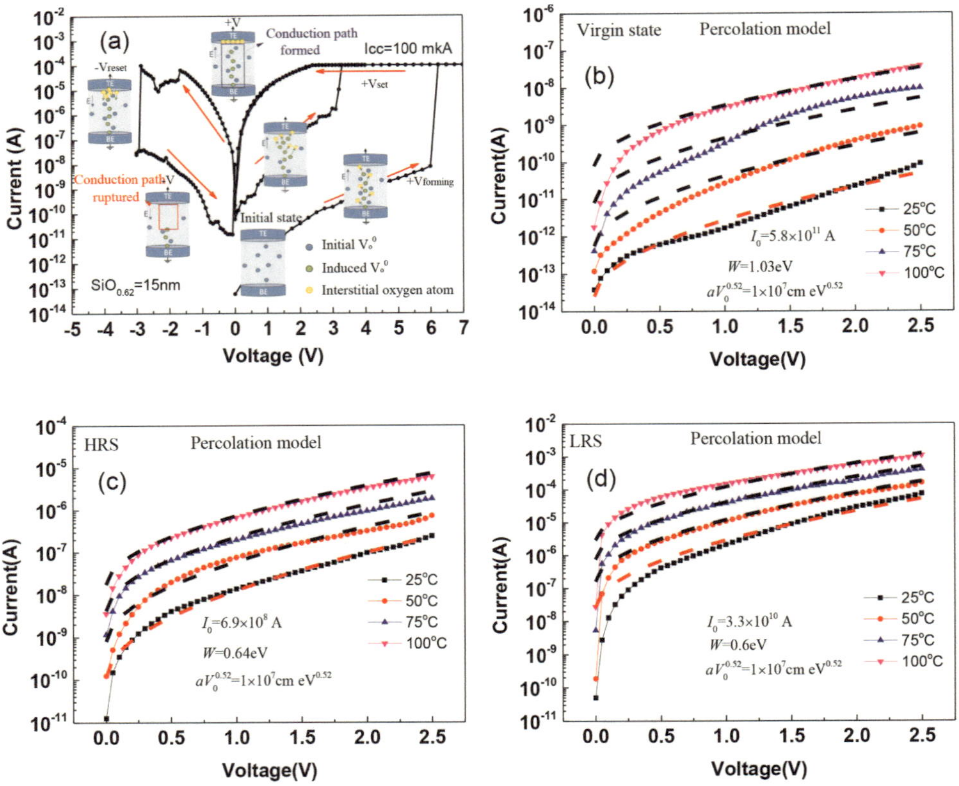

Figure 13.
(a) I-V *characteristics of* n^+*-Si/SiO*$_{0.62}$*/p*$^+$*-Si RRAM device under forming, set, and reset operations. I-V dependences of (b) VS, (c) HRS, and LRS (d) currents of* n^+*-Si/SiO*$_x$*/p*$^+$*-Si RRAM and fitting curves of Shklovskii-Efros model.*

HRS after a negative voltage bias. Then, the device was set to LRS again under a positive voltage bias. However, the positive set voltage was lower than the forming voltage once the RRAM switching function was established.

The current conduction mechanism is crucial for RRAM devices. To understand the conductive mechanism in this completely nonmetal RRAM, the measured *I-V* curves at different temperatures were further analyzed. **Figure 13(b)**, **(c)**, and **(d)** depict the measured and modeled *I-V* curves in the virgin state (VS), HRS, and LRS conditions, respectively. All state the HRS and LRS currents adhere to the Shklovskii-Efros percolation model [28]:

$$I = I_0 \exp\left(-\frac{W_e - \left(Ce\frac{U}{d}aV_0^{\gamma}\right)^{\frac{1}{1+\gamma}}}{kT}\right) \qquad (15)$$

where I_0, W_e, a, V_0, C, and γ are the preexponential factor, percolation energy, space scale of fluctuations, energy fluctuation amplitude, numeric constant which is equal to 0.25, and critical index which is equal to 0.9, respectively. The simulation by the Shklovskii-Efros model gives reasonable model parameters to all resistance state (**Figure 13(b–d)**). The percolation energy decreases with decreasing resistance. The relation $a \times {V_0}^{0.52} = 1 \times 10^{-7}$ cm eV$^{0.52}$ does not change from resistance to resistance. This is due to the fact that decreasing resistance increases space scale of fluctuations a, but decreases energy fluctuation amplitude V_0. In addition, it can be said that the Shklovskii-Efros percolation model is applicable to the LRS case,

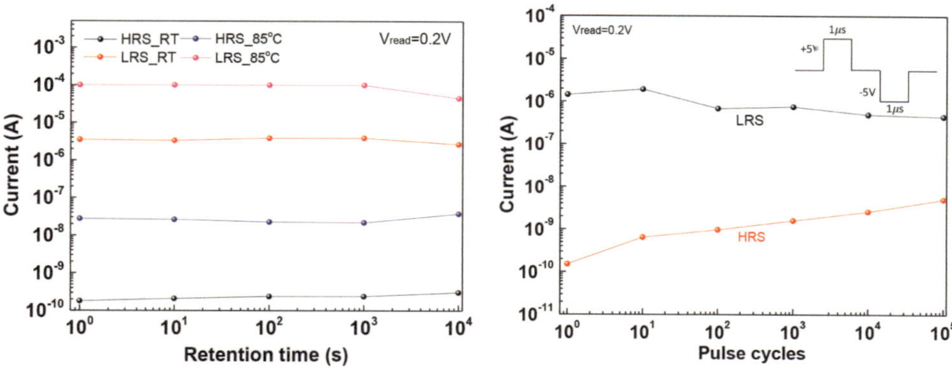

Figure 14.
(left) Retention characteristics of n^+-Si/$SiO_{0.62}$/p^+-Si RRAM devices at RT and 85°C. (right) Endurance characteristics of n^+-Si/$SiO_{0.62}$/p^+-Si RRAM devices.

and then, it can be assumed that the conducting channel is not continuous. Hence, the simulated results demonstrate that the charge transport of the n^+-Si/$SiO_{0.62}$/p^+-Si RRAM in VS, HRS, and LRS are described by the Shklovskii-Efros percolation model.

Figure 13(a) plots potential switching mechanisms. During the forming step, the current is conducted through the initial V_o^{2+} inside the SiO_x layer [31]. When the RRAM device was under sufficiently high positive voltage, soft breakdown in SiO_x occurred and disrupted the covalent bonds [32], generating unbonded Si ions, O^{2-}, and V_o^{2+}. It is assumed that after generation of anti-Frenkel pairs, electrons are redistributed to maintain charge neutrality, and new oxygen vacancies (V_o^0) and interstitial oxygen atoms are formed [30]. Because the atomic size of O is significantly smaller than Si, the interstitial oxygen atoms and V_o^0 could migrate inside SiO_x under the applied electric field. At the end of the forming process the interstitial oxygen atoms were attracted to the positive voltage and accumulated at the interface of top n^+-i junction. Once the conduction path was formed, electrons could transport through the V_o^0, creating the LRS current pass in the SiO_x layer. After application of a negative voltage, interstitial oxygen atoms moved away from the top n^+-i junction and recombined with V_o^0 to rupture the conduction path—the reset process. After a positive voltage was applied again, the set process behaved as the forming process to form a conduction path, but under a lower positive voltage than the forming voltage due to not all generating in the forming process V_o^0 recombined in the reset process.

Data retention and endurance are necessary characteristics for NVM, and they are related to the nonvolatile behavior and lifetime of an RRAM device. **Figure 14** (left) depicts the retention characteristics of the n^+-Si/$SiO_{0.62}$/p^+-Si RRAM device. The completely nonmetal RRAM device could achieve favorable retention with a slight resistive window decay from 1.9×10^4 to 8.7×10^3 at RT and 3.6×10^3 to 1.2×10^3 at 85°C after 10^4 s retention.

Figure 14 (right) depicts the pulsed endurance of the n^+-Si/$SiO_{0.62}$/p^+-Si RRAM device under set/reset pulses of +5/−5 V for 1 μs. In this case, higher voltages were used than DC switching cases because the energy to disrupt the covalent SiO_x bonds equals to the multiplication of *I, V,* and *time*. The resistance ratio between HRS and LRS decreased after increasing the pulsed cycles; however, the device exhibited excellent endurance with a resistance window of 89 after 10^5 pulsed switching cycles [33].

7. Conclusions

The silicon amorphous nanoclusters in as-deposited SiO_x and SiN_x films and silicon nanocrystals in the annealed films were studied using structural and optical methods. To analyze the sizes of silicon nanocrystals from the analysis of Raman scattering data, the phonon confinement model was refined.

From the analysis of XPS, Raman, and IR spectroscopy data, it has been established that pure random mixture and random bonding models do not adequately describe the real structure of the SiO_x and SiN_x films. The intermediate model was proposed. The nanoscale potential fluctuations in SiO_x and SiN_x films can be interpreted in the framework of the proposed model. The memristor effects in SiOx-based nonmetal structures were demonstrated.

Acknowledgements

The work was supported by the Russian Science Foundation (Project No. 18-49-08001) and the Ministry of Science and Technology (MOST) of Taiwan (No. 107-2923-E-009-001-MY3).

Author details

Vladimir Volodin[1,2*], Vladimir Gritsenko[1,2,3], Andrei Gismatulin[1] and Albert Chin[4]

1 Rzhanov Institute of Semiconductor Physics SB RAS, Novosibirsk, Russia

2 Novosibirsk State University, Novosibirsk, Russia

3 Novosibirsk State Technical University, Novosibirsk, Russia

4 National Chiao Tung University, Hsinchu, Taiwan, ROC

*Address all correspondence to: volodin@isp.nsc.ru

IntechOpen

© 2019 The Author(s). Licensee IntechOpen. This chapter is distributed under the terms of the Creative Commons Attribution License (http://creativecommons.org/licenses/by/3.0), which permits unrestricted use, distribution, and reproduction in any medium, provided the original work is properly cited.

References

[1] Pavesi L, Turan R, editors. Silicon Nanocrystals Fundamentals, Synthesis and Applications. Germany: Wiley; 2010. p. 652. ISBN: 978-3-527-32160-5

[2] Greben M, Khoroshyy P, Liu X, Pi X, Valenta J. Fully radiative relaxation of silicon nanocrystals in colloidal ensemble revealed by advanced treatment of decay kinetics. Journal of Applied Physics. 2017;**122**:034304. DOI: 10.1063/1.4993584

[3] Sychugov I, Juhasz R, Valenta J, Linnros J. Narrow luminescence linewidth of a silicon quantum dot. Physical Review Letters. 2005;**94**: 087405. DOI: 10.1103/PhysRevLett. 94.087405

[4] Park N-M, Kim T-S, Park S-J. Band gap engineering of amorphous silicon quantum dots for light-emitting diodes. Applied Physics Letters. 2001;**78**:2575-2577. DOI: 10.1063/1.1367277

[5] Anutgan T, Anutgan M, Atilgan I, Katircioglu B. Electroformed silicon nitride based light emitting memory device. Applied Physics Letters. 2017;**111**:053502. DOI: 10.1063/ 1.4997029

[6] Gismatulin AA, Kruchinin VN, Gritsenko VA, Prosvirin IP, Yen T-J, Chin A. Charge transport mechanism of high-resistive state in RRAM based on SiO_x. Applied Physics Letters. 2019;**114**: 033503. DOI: 10.1063/1.5074116

[7] Gritsenko VA. Atomic structure of amorphous nonstoichiometric silicon oxides and nitrides. Physics-Uspekhi. 2008;**178**:699-708. DOI: 10.1070/ PU2008v051n07ABEH006592

[8] Smith JE Jr, Brodsky MH, Crowder BI, Nathan MI. Raman spectra of amorphous Si and related tetrahedrally bonded semiconductors. Physical Review Letters. 1971;**26**:642-646. DOI: 10.1103/PhysRevLett.26.642

[9] Parker JH Jr, Feldman DW, Ashkin M. Raman scattering by silicon and germanium. Physical Review. 1967;**155**: 712-714. DOI: 10.1103/PhysRev.155.712

[10] Bustarret E, Hachicha MA, Brunel M. Experimental determination of the nanocrystalline volume fraction in silicon thin films from Raman spectroscopy. Applied Physics Letters. 1988;**52**:1675-1677. DOI: 10.1063/ 1.99054

[11] Zhigunov DM, Kamaev GN, Kashkarov PK, Volodin VA. On Raman scattering cross section ratio of crystalline and microcrystalline to amorphous silicon. Applied Physics Letters. 2018;**113**:023101. DOI: 10.1063/ 1.5037008

[12] Richter H, Wang ZP, Ley L. The one phonon Raman spectrum of microcrystalline silicon. Solid State Communications. 1981;**39**:625-629. DOI: 10.1016/0038-1098(81)90337-9

[13] Pailard V, Puech P, Laguna MA, Carles R, Kohn B, Huisken F. Improved one-phonon confinement model for an accurate size determination of silicon nanocrystals. Journal of Applied Physics. 1999;**86**:1921-1924. DOI: 10.1063/1.370988

[14] Volodin VA, Sachkov VA. Improved model of optical phonon confinement in silicon nanocrystals. Journal of Experimental and Theoretical Physics. 2013;**116**:87-94. DOI: 10.1134/ S1063776112130183

[15] Cardona M, Günterodt G, editors. Light Scattering in Solids II. Basic Concept and Instrumentation. Berlin: Springer-Verlag; 1982. p. 251. ISBN: 3-540-11380-0

[16] Cheng W, Ren S-F. Calculations on the size effects of Raman intensities of silicon quantum dots. Physical Review B. 2002;**65**:205305. DOI: 10.1103/PhysRevB.65.205305

[17] Keating PN. Effect of invariance requirements on the elastic strain energy of crystals with application to the diamond structure. Physics Review. 1966;**145**:637-645. DOI: 10.1103/PhysRev.145.637

[18] Dolling G. Lattice vibrations in crystals with the diamond structure. In: Inelastic Scattering of Neutrons in Solids and Liquids: II. Proceedings of IAEA, Vienna. 1963. pp. 37-48

[19] Kulda J, Strauch D, Pavone P, Ishii Y. Inelastic-neutron-scattering study of phonon eigenvectors and frequencies in Si. Physical Review B. 1994;**50**:13347-13356. DOI: 10.1103/PhysRevB.50.13347

[20] Kirk CT. Quantitative analysis of the effect of disorder-induced mode coupling on infrared absorption in silica. Physical Review B. 1988;**38**:1255-1273. DOI: 10.1103/PhysRevB.38.1255

[21] Pai PG, Chao SS, Takagi Y, Lucovsky G. Infrared spectroscopic study of SiO_x films produced by plasma enhanced chemical vapor deposition. Journal of Vacuum Science and Technology A. 1986;**4**:689-694. DOI: 10.1116/1.573833

[22] Yin Z, Smith FW. Optical dielectric function and infrared absorption of hydrogenated amorphous silicon nitride films: Experimental results and effective-medium-approximation analysis. Physical Review B. 1990;**42**:3666-3675. DOI: 10.1103/PhysRevB.42.3666

[23] Korchagina TT, Marin DV, Volodin VA, Popov AA, Vergnat M. Structure and optical properties of SiN_x: H films with Si nanoclusters produced by low frequency plasma enhanced chemical vapor deposition. Semiconductors. 2009;**43**:1514-1520. DOI: 10.1134/S1063782609110207

[24] Volodin VA, Bugaev KO, Gutakovsky AK, Fedina LI, Neklyudova MA, Latyshev AV, et al. Evolution of silicon nanoclusters and hydrogen in SiN_x:H films: Influence of high hydrostatic pressure under annealing. Thin Solid Films. 2012;**520**:6207-6214. DOI: 10.1016/j.tsf.2012.05.019

[25] Gritsenko VA, Xu JB, Kwok RWM, Ng YN, Wilson IH. Short range order and the nature of defects and traps in amorphous silicon oxynitride governed by the Mott rule. Physical Review Letters. 1998;**81**:1054-1057. DOI: 10.1103/PhysRevLett.81.1054

[26] Bell FG, Ley L. Photoemission study of SiO_x ($0 \leq x \leq 2$) alloys. Physical Review B. 1988;**37**:8383-8393. DOI: 10.1103/PhysRevB.37.8383

[27] Hickmott TW, Baglin JE. Stoichiometry and atomic defects in rf-sputtered SiO_2. Journal of Applied Physics. 1979;**50**:317-323. DOI: 10.1063/1.325662

[28] Shklovskii BI, Efros AL. Electronic Properties of Doped Semiconductors. Heidelberg: Springer; 1984. p. 416

[29] Strukov DB, Snider GS, Stewart DR, Williams RS. The missing memristor found. Nature. 2008;**453**:80-83. DOI: 10.1038/nature06932

[30] Mehonic A, Shluger AL, Gao D, Valov I, Miranda E, Ielmini D, et al. Silicon oxide (SiO_x): A promising material for resistance switching? Advanced Materials. 2018;**30**:1801187. DOI: 10.1002/adma.201801187

[31] Dong LP, Jia RX, Xin B, Peng B, Zhang YM. Effects of oxygen vacancies

on the structural and optical properties of β-Ga_2O_3. Scientific Reports. 2017;**7**: 40160. DOI: 10.1038/srep40160

[32] Padovani A, Gao DZ, Shluger AL, Larcher L. A microscopic mechanism of dielectric breakdown in SiO_2 films: An insight from multi-scale modeling. Journal of Applied Physics. 2017;**121**: 155101. DOI: 10.1063/1.4979915

[33] Te JY, Gismatulin A, Volodin V, Gritsenko V, Chin A. Novel all nonmetal resistive random access memory. Scientific Reports. 2019;**9**(1):6144. DOI: 10.1038/s41598-019-42706-9

Chapter 3

Ferrite-Based Nanoparticles Synthesized from Natural Iron Sand as the Fe^{3+} Ion Source

Malik Anjelh Baqiya, Retno Asih, Muhammad Ghufron, Mastuki, Dwi Yuli Retnowati, Triwikantoro and Darminto

Abstract

Ferrite-based nanoparticles, namely, bismuth ferrite ($BiFeO_3$) and calcium ferrite ($CaFe_4O_7$), have been synthesized via sol-gel and chemically dissolved method, respectively, employing hematite (α-Fe_2O_3) as the Fe^{3+} ion source. Firstly, α-Fe_2O_3 nanoparticles were prepared from natural iron sand containing mostly magnetite (Fe_3O_4) phase through coprecipitation technique continued by sintering process at 800°C for 2 h. Higher $BiFeO_3$ phase content was achieved after Bi-Fe gel being annealed at 650°C for 1 h in air atmosphere. Furthermore, major phase of $CaFe_4O_7$ was formed with molar ratio of Fe^{3+}/Ca^{2+} = 6 and sintering temperature of 800°C for 3 h. Interestingly, the powders with dominant $CaFe_4O_7$ phase, known as calcium biferrite, exhibit higher ferromagnetism at room temperature. The magnetic properties of the calcium biferrite are comparable to those of barium hexaferrite which can be applied for radar-absorbing material. Meanwhile, $BiFeO_3$ powders also show weak room temperature ferromagnetism. It has also demonstrated that Ni doping in the bismuth ferrite ($BiFe_{1-x}Ni_xO_3$ with x = 0.1) nanoparticles results in enhancement of the magnetic properties. Moreover, a ferroelectric hysteresis loop and a trend of frequency dependence of the dielectric constant have been observed, which were enhanced by Pb doping ($Bi_{1-y}Pb_yFeO_3$ with y = 0.1). These results suggest a multiferroic behavior in the $BiFeO_3$ nanoparticles.

Keywords: bismuth ferrite, calcium ferrite, iron sand, multiferroic, nanoparticles, precipitation, sol-gel

1. Introduction

Development of functional nanomaterials for scientific and industrial applications is very crucial for advanced technologies. The use of natural resources as the starting compounds for producing nanomaterials is currently developing. Many researchers are exploring natural materials and even waste biomass applied as a functional material that has a high selling value for various specific applications. For example, the use of silica sand from Tanah Laut, Kalimantan, Indonesia, as a raw material for manufacturing pure SiO_2, zircon, and zirconia with high phase purity and crystallite size in nanometer range was reported [1]. Moreover, natural

IntechOpen

iron sand exploration as a starting material has been shown to produce magnetite (Fe_3O_4) nanoparticles as magnetic coating, magnetic fluid (ferrofluid), and magnetic gel (ferrogel) for radar-absorbing materials, biomedical applications, and tissue engineering, respectively [2–5].

Fe_3O_4 is one of the magnetic particles that can be obtained from natural iron sand after conducting the separation technique from its impurities by mechanical and chemical processes. In nature, iron sand consists of more than 90 wt% of Fe_3O_4 particles. Generally, Fe_3O_4 has been synthesized using commercial raw materials, such as $FeCl_2.4H_2O$ and $FeCl_3.6H_2O$ [6]. The commonly used synthesis methods are sol-gel, hydrothermal, and coprecipitation techniques [7–9]. Because Fe_3O_4 nanoparticles tend to agglomerate among particles, the addition of surfactants or templates has been widely applied to produce homogeneous nanoparticles with certain sizes and morphologies [10–14]. Research on preparing Fe_3O_4 nanoparticles from iron sand has been the main topic for the past few years. The use of doping, for example, doping Mn and Zn, on Fe_3O_4 makes it superparamagnetic so that it can be applied in biomedicine applications [15–18].

Hematite (α-Fe_2O_3) is the most stable iron oxides at high temperatures. α-Fe_2O_3 is commonly obtained from iron rust which is one of the dominant corrosion products of iron metal or iron alloys. In general, α-Fe_2O_3 nanoparticles have been successfully prepared by several methods, namely, hydrothermal [19] and coprecipitation technique [20], using commercial raw materials, such as $Fe(NO_3)_3{\cdot}9H_2O$ and $FeCl_3.6H_2O$, respectively. It is found that the concentration of Fe^{3+} ions used in the preparation of α-Fe_2O_3 nanoparticles may influence the particle size and morphology, as well as the optical bandgap [20]. α-Fe_2O_3 nanoparticles with particle size of 8 nm possess superparamagnetic properties with relatively high magnetization at room temperature [21]. Therefore, it is possible to be applied for biomedical and spintronic applications. Moreover, Liu et al. have successfully prepared porous Fe_2O_3 nanorods with particle size of ~10 nm and pore sizes in the range of 5–50 nm. These porous Fe_2O_3 nanorods exhibit excellent photocatalytic properties [22].

In the field of environmental engineering, α-Fe_2O_3 nanoparticles can be synthesized from hydrated ferric chloride and ferrous sulfate salt solution through chemical coprecipitation method and calcination process at relatively high temperature of 500°C [23]. In addition, a simple chemically coprecipitation method has been employed to obtain Fe_3O_4 nanoparticles using HCl and NH_4OH as dissolving and precipitating agent, respectively [3]. Some researchers have investigated the transformation from Fe_3O_4 to α-Fe_2O_3 phase through oxidation process of Fe^{2+} to Fe^{3+} ions [24]. It is noted that Fe_3O_4 nanoparticles could be transformed into maghemite (γ-Fe_2O_3) and hematite (α-Fe_2O_3) via dry oxidation process at temperature range between 350 and 400°C and 600 and 800°C, respectively [25]. Focusing on the use of natural resources as raw materials for synthesizing functional materials, in this chapter, α-Fe_2O_3 nanoparticles were synthesized from natural iron sand through chemical coprecipitation method followed by sintering process at temperature of 800°C. Then, the obtained α-Fe_2O_3 nanoparticles were utilized as one of the raw materials for preparing calcium ferrite (Ca-ferrites) and bismuth ferrite ($BiFeO_3$) nanoparticles as potential materials for radar-absorbing and data storage materials, respectively. The physical characterizations for all obtained ferrite-based nanoparticles include elemental and phase identification, particle morphology, and magnetic and electrical properties.

Based on the phase diagram of CaO-Fe_2O_3 system [26, 27], it is known that there are three main phases of calcium ferrite compounds and those are $2CaO.Fe_2O_3$ ($Ca_2Fe_2O_5$), $CaO.Fe_2O_3$ ($CaFe_2O_4$), and $CaO.2Fe_2O_3$ ($CaFe_4O_7$). It is possible that the reaction between CaO and Fe_2O_3 results in other unstable calcium ferrite phases, such as $CaFe_{12}O_{19}$. In addition, Boyanov [28] has pointed out that the mixture of

$CaCO_3$-Fe_2O_3 after thermal treatment has produced various types of calcium ferrite compounds consisting of ~50% $CaO.2Fe_2O_3$, ~20% $CaO.Fe_2O_3$, ~8% $2CaO.Fe_2O_3$, and other ferrite products. The formation of calcium ferrite compounds depends on the kinetics of chemical reaction at the boundary between the phases and oxide diffusion during the reaction affected by the concentration ratio of the existing Ca^{2+} and Fe^{3+} ions as the precursors and also the atmospheric condition [29].

Calcium ferrite compounds exhibit soft ferromagnetism, and, therefore, it can be used for radar-absorbing materials in the calcium ferrite/graphite nanocomposites [30]. In this case, calcium ferrite nanoparticles have magnetic properties that are comparable to barium ferrite ($BaO.6Fe_2O_3$) and strontium ferrite ($SrO.6Fe_2O_3$) known as M-type hexaferrite for microwave-absorbing applications. In order to be used for this application and also for biomedical applications as targeted drug delivery, calcium ferrite should exhibit superparamagnetic behavior [31]. Compared with the other ferrites, such as MFe_2O_4 (M = Zn, Mn, Ni, and Cu), $CaFe_2O_4$ is one of the biocompatible materials and environmentally friendly due to the use of calcium rather than heavy metals. Moreover, $Ca_2Fe_2O_5$ with the brownmillerite structure has a specific application as p-type thermoelectric device [32]. This is due to the fact that this compound has interesting electrical properties [33, 34]. Oxygen deficiencies in the $Ca_2Fe_2O_5$ crystals may enhance the electrochemical activity [35]. On the other hand, $CaFe_4O_7$ has not been explored yet regarding its magnetic properties. In contrast to the other calcium ferrites, in this chapter, $CaFe_4O_7$ nanoparticles were prepared by mixing Fe_2O_3 from natural iron sand and $CaCO_3$ from natural limestone.

Bismuth ferrite ($BiFeO_3$) is one of multiferroic system showing a magnetic-electric coupling at room temperature. Multiferroic material has perovskite structure with chemical formula ABO_3. The type of A and B sites, the cation nonstoichiometry, and the presence of oxygen vacancies may have an impact on the structural, electronic, and magnetic properties [36]. $BiFeO_3$ crystallizes in a distorted rhombohedral perovskite with space group R3c [37]. It has high Curie temperature and Néel temperature of 1100 and 640 K, respectively [38]. It is difficult to obtain a pure phase of $BiFeO_3$ because the kinetics of phase formation leads to the formation of secondary phases, such as $Bi_{25}FeO_{40}$ (sillenite) and $Bi_2Fe_4O_9$ (mullite). Various techniques have been reported to prepare single phase of $BiFeO_3$, and those are chemical coprecipitation [39], hydrothermal [40], and sol-gel methods [41–43]. The ideas of those techniques are to achieve a single phase of $BiFeO_3$ with a simple route, low temperature, and cost-effectiveness. Wang et al. have found that the formation of $BiFeO_3$ phase starts at 425°C with impurity phases about 30% by the low-heating temperature solid-state precursor method [44, 45]. Further calcination from 450 to 550°C results in a pure $BiFeO_3$ phase without any impurity phases. However, impurity phase of $Bi_2Fe_4O_9$ has been detected in the powder calcined at above 650°C. Moreover, $BiFeO_3$ nanoparticles synthesized by microwave-assisted sol-gel method at calcination temperature of 450°C exhibit a pure phase of $BiFeO_3$ structure with particle size of 40 nm and no detected secondary phase [46].

Magnetic and dielectric properties of $BiFeO_3$ nanoparticles are determined by the introduction of doping and particle size influenced by the synthesis method, temperature, and duration of calcination. It has been found that all magnetic parameters, such as saturation magnetization, enhance with decreasing particle size [43]. $BiFeO_3$ nanoparticles with the size below 100 nm have weak ferromagnetism at room temperature. This ferromagnetic behavior in the nanoparticles is due to the presence of oxygen vacancies in $BiFeO_3$ system [41, 47]. Enhancement of magnetic as well as dielectric properties in $BiFeO_3$ can be achieved by adding doping of Mn, Ni, Pb, Ti, Sr, and Zn [48–56]. Up to the present, there have been various studies examining the doping effects of $BiFeO_3$ nanoparticles with numerous advanced techniques to improve their performance. In the case of the enhancing magnetization induced by doping, it has been suggested that this is probably due to increasing distortion of local structure, increasing the effect of Dzyaloshinskii-Moriya (DM) interaction, distortion of Fe and O bonding, destruction of spin cycloid structure, and the presence of impurity phase in the $BiFeO_3$ systems [53, 57]. Besides affecting the magnetic properties, introduction of doping in $BiFeO_3$ leads to the improvement of dielectric and ferroelectric properties [50, 58, 59]. Yuan et al. [54] have found that a sufficient amount of Sr/Pb doping can improve the magnetic properties as well as high-frequency dielectric properties.

In addition, the dielectric properties of pure $BiFeO_3$ phase strongly depend on the atmospheric condition during the powder synthesis. Liu et al. [60] have found a higher spontaneous polarization and lower breakdown field based on polarization-electrical field (P-E) hysteresis loops in the samples annealed in H_2 and N_2 atmospheres. In this chapter, $BiFeO_3$ nanoparticles were synthesized by sol-gel method using natural iron sand as one of the raw materials and calcined in air atmosphere. Then, the ferroelectric and the dielectric properties were intensively investigated in the Pb- and Ni-doped $BiFeO_3$ nanoparticles.

2. Preparation of hematite (α-Fe_2O_3) nanoparticles

Prior to the preparation of α-Fe_2O_3 nanoparticles, at first, Fe_3O_4 nanoparticles were synthesized from natural iron sand as the raw material by coprecipitation technique using HCl as dissolving agent and NH_4OH as precipitating agent. The detail of experimental procedure to synthesize Fe_3O_4 nanoparticles was also described in elsewhere [3]. First of all, the extracted iron sand was collected and dissolved in 12 M HCl at ~70°C under continuous and constant stirring of 600 rpm. The obtained solution from the reaction process was filtered and added slowly with 6.49 M NH_4OH under the same temperature and stirring speed for 30 minutes. Then, the black precipitates were formed. The precipitate (Fe_3O_4 phase) was initially washed with distilled water until pH 7 and then dried at 70°C for 5 h. In order to get α-Fe_2O_3 phase, the dried nanopowder (Fe_3O_4 phase) was calcined at 800°C for 2 h, as shown in **Figure 1**. Finally, the Fe_2O_3 powders from this calcination were continued by performing coprecipitation process again with the same experimental procedure as before until the precipitation process. A reddish

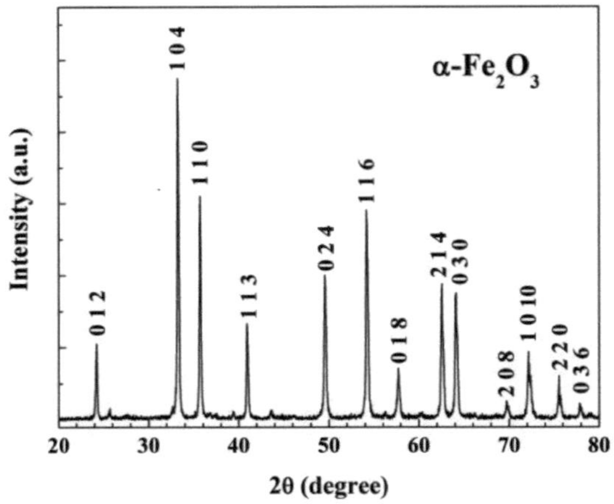

Figure 1.
Hematite (α-Fe_2O_3) synthesized from natural iron sand (Fe_3O_4) by coprecipitation method followed by calcination process at 800°C for 2 h.

precipitate ($Fe_2O_3.H_2O$) was formed. The resulted precipitate was then washed and collected for further synthesis of $CaFe_4O_7$ and $BiFeO_3$ (without and with doping of Pb and Ni) nanoparticles.

3. Preparation of calcium ferrite nanoparticles

Calcium biferrite ($CaFe_4O_7$) nanoparticles were synthesized by the so-called chemically dissolved method using precipitated $CaCO_3$ and Fe_2O_3 as Ca^{2+} and Fe^{3+} ion sources, respectively. Fe_2O_3 powders were obtained as described previously from natural iron sand, whereas the precipitated $CaCO_3$ particles were synthesized from natural limestone through carbonation process. First, the natural limestone was extracted from the existing impurities, such as silica, and then it was calcined at 900°C for 6 h to produce CaO. The CaO powder was dissolved into distilled water to produce $Ca(OH)_2$ solution. The carbonation process using CO_2 gas flow was performed until it formed a precipitation at pH around 7. The precipitated $CaCO_3$ was filtered and dried for further synthesis. The detail procedure was also explained in the former paper by Arifin et al. [61].

In the synthesis of the calcium ferrite nanoparticles using the chemically dissolved method, the obtained Fe_2O_3 and precipitated $CaCO_3$ were dissolved in HNO_3 to get $Fe(NO)_3$ and $Ca(NO)_2$ solutions, respectively, with a molar ratio of 1:6. Both solutions were mixed homogeneously and heated at constant temperature (80°C) and stirring rate (600 rpm) until it formed slurry precipitates. The precipitates were washed using distilled water and dried at 80°C for 10 h. The resulted powders were collected and then sintered at 800°C for 3 h.

4. Preparation of bismuth ferrite ($BiFeO_3$) nanoparticles without and with Pb and Ni doping

Nanoparticles of undoped, Pb- and Ni-doped $BiFeO_3$ ($BiFeO_3$, $Bi_{0.9}Pb_{0.1}FeO_3$, and $BiFe_{0.9}Ni_{0.1}O_3$, respectively) were prepared by sol-gel method. The starting materials were Fe_2O_3 synthesized previously from iron sand (94%) as the Fe^{3+} ion source and Bi_2O_3 (Aldrich, 99.9%) as the Bi^{3+} ion source. $Pb(NO_3)_2$ (powder, 99%) and $Ni(NO_3)_2.6H_2O$ (powder, 99%) were used as the Pb and Ni doping, respectively. Fe_2O_3, Bi_2O_3, $Pb(NO_3)_2$, and $Ni(NO_3)_2.6H_2O$ powders were dissolved separately by HNO_3 (Merck, 65%) to form solutions of ferrite nitrate, bismuth nitrate, lead nitrate, and nickel nitrate, respectively, with the stoichiometric molar ratio of (Bi, Pb):(Fe, Ni) = 1:1. Acetic acid was added into each solution under constant stirring and temperature for 30 minutes. Then, it was followed by addition of ethylene glycol under the same condition. Next, the obtained solutions were mixed together under the same temperature and stirring rate for 1 h. The resulted solution was dried at 80°C for 6 days to obtain the undoped and doped $BiFeO_3$ xerogels. The dried gels were ground and collected. Finally, the powders were calcined in air at 650 and 700°C for 1 h to form undoped $BiFeO_3$ and doped $BiFeO_3$ ($Bi_{0.9}Pb_{0.1}FeO_3$ and $BiFe_{0.9}Ni_{0.1}O_3$), respectively, for further characterizations.

5. Characterizations

A thermogravimetric/differential thermal analysis (TG/DTA) was performed to determine the thermal behaviors of the dried gel of bismuth ferrite. The phase formation and crystal structure of all samples were characterized by X-ray diffraction (XRD) with Cu-K_α radiation and λ = 1.54056 Å for scanning 2θ range of 20–70°. The lattice parameters and average crystallite sizes were determined by XRD patterns which were analyzed by the Rietveld method using the Rietica and MAUD programs [62, 63]. Transmission electron microscopy (TEM) with selected area electron diffraction (SAED) pattern was conducted to investigate the particles' morphology and crystal structure confirmation of all ferrite-based samples. The magnetic properties of the nanoparticles were measured using vibrating sample magnetometry (VSM, Oxford VSM1.2H) and superconducting quantum interference device (SQUID) magnetometer in external magnetic field range of ±1 T at room temperature. The ferroelectric properties of the bismuth ferrites were studied from the polarization-electric field (P-E) hysteresis loops using a polarization meter (Radiant Technologies 66A). Frequency dependence of the dielectric constant of all bismuth ferrites was estimated by two-probe electrical resistance using Automatic RCL Meter (type PM6303A).

6. Structural and magnetic properties of calcium ferrites from natural iron sand and limestone

Figure 2 shows the XRD pattern of calcium ferrite compound synthesized by the chemically dissolved method from natural iron sand and limestone as the raw materials and then sintered at 800°C for 3 h. Based on the analysis of phase identification, it can be seen that the resulted powder contains several phases of calcium ferrites, $CaFe_4O_7$, $Ca_4Fe_{14}O_{25}$, and $Ca_2Fe_9O_{13}$, with weight percentages of 28.8, 46.6, and 24.6 wt%, respectively. The formation of those phases is possible to occur due to the atmospheric condition during calcination. Generally, at relatively high calcination temperatures, the most stable phases are those that have higher coordination numbers, in this case with surrounding oxygen. Hughes et al. [64] have also identified these distinct calcium ferrite phases in the mixture of CaO and Fe_2O_3 calcined in air at high temperatures between 1180 and 1240°C. In addition, the phase formation of $Ca_2Fe_9O_{13}$ can be present in the compound at the lower temperatures [65]. With the increase of temperature, the phase formation becomes more complex. Related to the phase transformation, it strongly depends on the crystallization kinetics of the reaction, the ratio concentration between Ca and Fe ions, and the atmospheric condition [66].

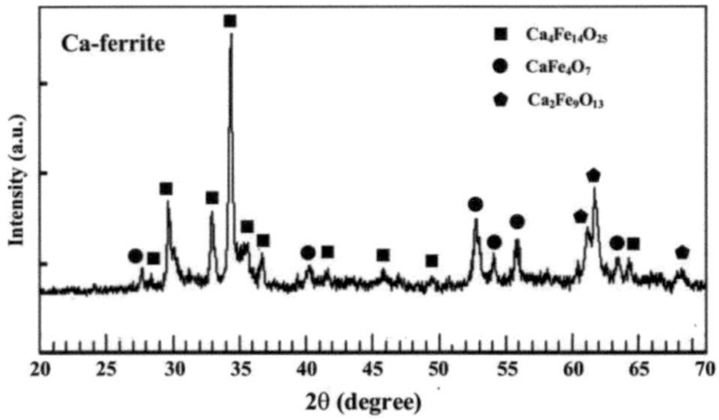

Figure 2.
XRD pattern of calcium ferrite powders synthesized by the chemically dissolved technique from natural iron sand and limestone as the Fe^{3+} and Ca^{2+} ion sources, respectively, and then continued by calcination process at 800°C for 3 h.

Focusing on the high intensities of the diffraction peaks, the sample exhibits XRD lines of both $CaFe_4O_7$ and $Ca_4Fe_{14}O_{25}$ phases as the dominant phases. $CaFe_4O_7$ has monoclinic structure and $Ca_4Fe_{14}O_{25}$ has hexagonal structure. Both phases have similar crystalline structure related to hexagonal ferrite structures [67]. The XRD pattern in **Figure 2** shows that $CaFe_4O_7$ and $Ca_4Fe_{14}O_{25}$ phases have broad diffraction peaks. This indicates that the average crystallite sizes are in a nanometer scale. Based on the Rietveld analysis, $CaFe_4O_7$ phase in the calcium ferrite compound has average crystallite size of about 46 nm. In order to clarify the nano-sized particles, TEM image is important to be investigated in detail.

Figure 3 displays TEM image of the calcium ferrite sample together with the selected area electron diffraction (SAED). The TEM image proves that the particle size of the sample is in the range of 40–60 nm. This is in a good agreement with the Rietveld analysis of the XRD pattern in **Figure 2**. The analysis of electron diffraction from SAED pattern reveals that $CaFe_4O_7$ and $Ca_4Fe_{14}O_{25}$ phases are dominantly present and $Ca_2Fe_9O_{13}$ is the minor phase in the sample. This result is also consistent with the XRD pattern analysis.

Magnetic properties of the calcium ferrite compound were studied by the magnetic hysteresis curve (M-H curve) at room temperature as shown in **Figure 4**. It is clear that the sample exhibits ferromagnetic behavior. A detailed observation on the M-H curve of the sample shows that the values of remanent magnetization and magnetization at 1 T are 2.11 and 10.94 emu/g, respectively. This indicates that a soft magnetism is realized in the calcium ferrite compound. It has been found that the dominant phase existing in the sample has a contribution to the ferromagnetic behavior [68]. The value of magnetism in the sample is comparable with that of the barium-calcium hexaferrite prepared by sol-gel and microemulsion techniques, in which the saturation magnetization value is approximately 24 emu/g [69]. Moreover, Samariya et al. [70] have studied the magnetic properties of calcium ferrite, in the form of $CaFe_2O_4$, nanoparticles. They have found similar value of magnetization compared with the present result in this work. Concerning the multiphase compound, the magnetic parameters in the sample are influenced by the presence of nonmagnetic phase, magnetic domain and its orientation, and defect formation. Therefore, it is important to investigate more detail on how to prepare a pure certain phase of calcium ferrite from natural resources as the starting materials. Accordingly, this result demonstrates that the present calcium ferrite nanoparticles could be used as one of the potential materials for microwave absorption application.

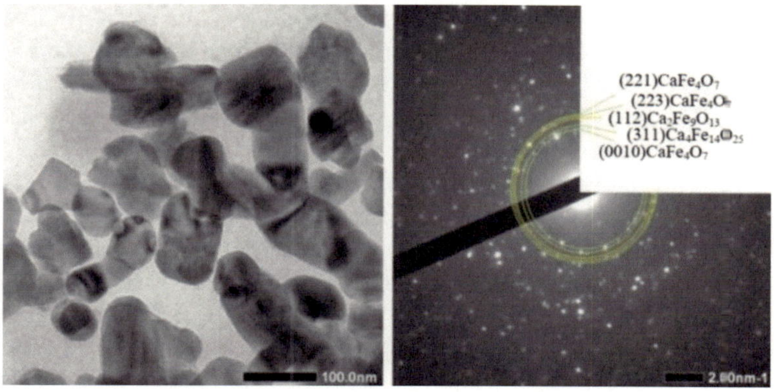

Figure 3.
TEM image with selected area electron diffraction (SAED) pattern for calcium ferrite powders synthesized by the chemically dissolved technique from natural iron sand and limestone as the Fe^{3+} and Ca^{2+} ion sources, respectively, and then continued by calcination process at 800°C for 3 h.

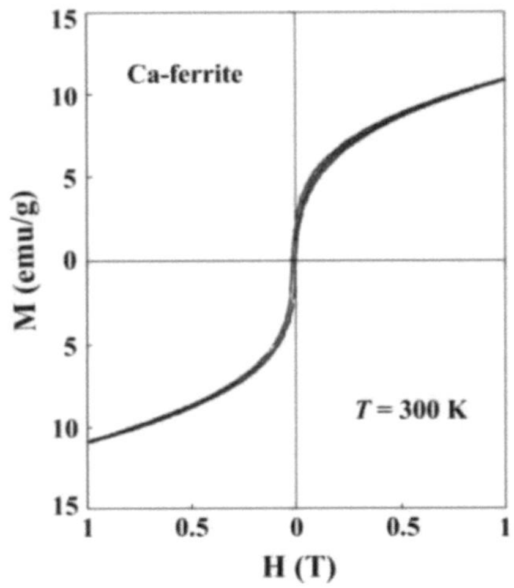

Figure 4.
Magnetization curve at room temperature for calcium ferrite powders synthesized by the chemically dissolved technique from natural iron sand and limestone as the Fe^{3+} and Ca^{2+} ion sources, respectively, and then continued by calcination process at 800°C for 3 h.

7. Magnetoelectric properties of bismuth ferrite nanoparticles

TG/DTA curve of the uncalcined powder of the undoped $BiFeO_3$, shown in **Figure 5**, exhibits about 29% weight loss from room temperature to 550°C due to the evaporation of water, organics, and nitrate decomposition [71, 72]. Based on this thermal behavior, the powder could be thermally treated at temperatures from 500 up to 700°C for 1 h. Carvalho et al. [73] have reported that the increasing time of the heat treatment increases the formation of secondary phases and, therefore, they have suggested to avoid a long heat treatment to synthesize $BiFeO_3$ nanoparticles.

Figure 6 shows the XRD patterns of the undoped and doped $BiFeO_3$ samples calcined at 650 and 700°C, respectively, for an hour in air atmosphere. This heat treatment was conducted to form $BiFeO_3$ phase. The influence of the atmosphere in the phase formation has been investigated by Xu et al. [72]. They have reported that crystallization in the atmosphere is important to obtain a pure $BiFeO_3$ phase prepared by sol-gel method. It can be seen from the phase identification of the XRD patterns that multiphases of bismuth ferrite compounds such as $BiFeO_3$, $Bi_{25}FeO_{40}$, and $Bi_2Fe_4O_9$ were observed in the synthesized powders. Moreover, Bi_2O_3 was still observed in the XRD patterns in minor composition. $BiFeO_3$ is a metastable phase which easily decomposes to secondary phases, $Bi_{25}FeO_{40}$ and $Bi_2Fe_4O_9$, at high temperatures [73]. In this present work, it is found that higher $BiFeO_3$ phase is achieved with heat treatment at 650°C for 1 h. This result is consistent with the TG/DTA and XRD data analyzed by Sakar et al. [74] which corresponds to sharp diffraction peaks of the $BiFeO_3$ phase. The formation of secondary phases increases at higher temperature than 650°C. $BiFeO_3$ began to decompose because of its unstable thermodynamic character when the calcination temperature was further increased. The relative weight percent and average crystallite size of the $BiFeO_3$ phase were determined from the diffraction patterns by Rietveld method using Rietica and MAUD program, respectively. Overall, the analysis results show that the bismuth ferrite powders contain about 75 wt% of $BiFeO_3$ phase. The average crystallite size of the $BiFeO_3$ sample prepared at 650°C is about 84 nm.

The addition of doping substituting the A and B sites in the ABO_3 perovskite structure of $BiFeO_3$ greatly affects the crystal distortion and changes in the

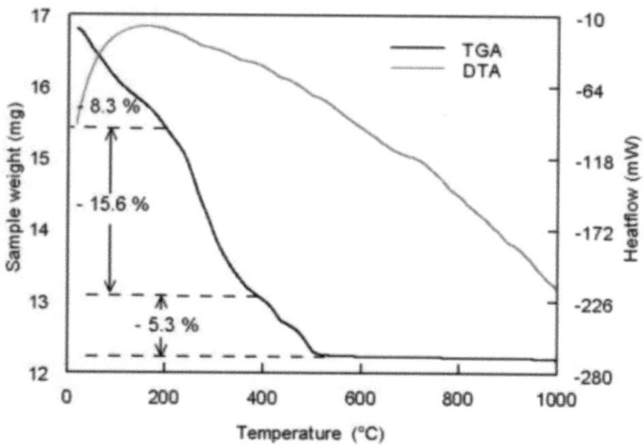

Figure 5.
TG/DTA curves of the uncalcined $BiFeO_3$ powder.

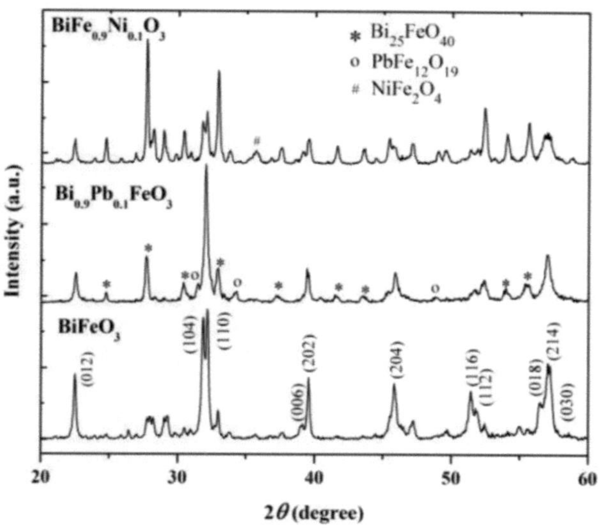

Figure 6.
XRD patterns of the undoped $BiFeO_3$ and doped $BiFeO_3$ ($Bi_{0.9}Pb_{0.1}FeO_3$ and $BiFe_{0.9}Ni_{0.1}O_3$) powders synthesized by sol-gel method calcined at 650 and 700°C, respectively, for 1 h in air.

composition of the secondary bismuth ferrite phases. Pb ion substitutes A site, namely, the Bi^{3+} ion, in the structure of $BiFeO_3$. As a result, Pb doping has an effect on the diffraction peak shift of the $BiFeO_3$ phase to the lower diffraction angle. This is because the ionic radius of Pb^{2+} ion (0.119 nm) is greater than that of Bi^{3+} ion (0.103 nm). Moreover, it can also be seen that there is a combination of the diffraction peaks for the crystal plane (006) and (202) into the diffraction peak (111) at 2θ of 31–32°. This indicates a small change in the distortion of the crystal from distorted rhombohedral to pseudocubic system. XRD analysis confirms that $Bi_{0.9}Pb_{0.1}FeO_3$ has cubic structure with space group of *Pm*-3 *m*, compared with the undoped $BiFeO_3$ having rhombohedral structure with space group of R3c. It is important to mention that the secondary phase in the Pb-doped $BiFeO_3$ ($Bi_{0.9}Pb_{0.1}FeO_3$) sample, which is $PbFe_{12}O_{19}$, has been reported to be one of the hexaferrite materials exhibiting good superparamagnetic behavior [75]. Further Rietveld analysis from the XRD patterns gives the values of lattice parameters of $BiFeO_3$, $Bi_{0.9}Pb_{0.1}FeO_3$, and $BiFe_{0.9}Ni_{0.1}O_3$ as shown in **Table 1**.

On the XRD pattern of the Ni-doped $BiFeO_3$ ($BiFe_{0.9}Ni_{0.1}O_3$) sample, shown in **Figure 6**, it is clear that there is no change of the crystal structure due to Ni doping at the B site (Fe^{3+} ion) of $BiFeO_3$ crystal. This is displayed by the rhombohedral peak which can still be observed at 2θ of 31–32°. The result of the phase composition analysis gives that there is an increase of secondary phases ($Bi_{25}FeO_{40}$) and the presence of $NiFe_2O_4$ in the sample. Interestingly, both secondary phases have also unique magnetoelectric properties. It has been reported by Zhu et al. [76] that $Bi_{25}FeO_{40}$ has good dielectric and electrical properties which can be used as one of integrated circuit components. $NiFe_2O_4$ is one of magnetic spinel structures with good magnetic and dielectric properties [77]. In addition, Ni doping in the $BiFeO_3$ system has an effect on diffraction peak shift to the lower diffraction angle because ionic radius of Ni^{3+} ion (0.069 nm) is slightly larger than that of Fe^{3+} ion (0.065 nm). The change of lattice parameter due to Pb and Ni doping in $BiFeO_3$ system is summarized in **Table 1**.

Figure 7 shows the TEM image and selected area electron diffraction (SAED) patterns of $BiFeO_3$ powders annealed at 650°C for 1 h in air. Sharp diffraction spots seen from SAED pattern confirm the formation of well crystalline bismuth ferrites. Phases identified from SAED pattern are relatively matching with the XRD patterns in **Figure 6** consisting of $BiFeO_3$, $Bi_{25}FeO_{40}$, $Bi_2Fe_4O_9$, and Bi_2O_3. The TEM image shows typical morphology of particle agglomeration. The particle size is greater than the average crystallite size estimated by Rietveld analysis due to agglomeration of the nanoparticles.

The nonlinear magnetic hysteresis curve of the bismuth ferrite powders, as shown in **Figure 8**, illustrates weak ferromagnetism. The remanent magnetization of 0.044 emu/g and coercive field of 68.5 Oe in the undoped $BiFeO_3$ confirm the weak ferromagnetism behavior at room temperature. The complete saturation of magnetization of powders was not achieved up to applied magnetic field of 1 T. The hysteresis loop of bulk $BiFeO_3$ is generally linear indicating antiferromagnetic order at the ground state (5 K) [78]. The weak ferromagnetic order of these powders can be understood as a result of residual magnetic moment caused by its canted spin structure [79]. The canting of the spins can be caused by reduction of particle size. When the particle size decreases, the number of surface asymmetry atoms increases, then it changes the angle of the helical ordered spin arrangement, and finally the net magnetic moment appears [80]. Moreover, the existence of defects, for instance, oxygen vacancies [81], and the secondary phases [82] may contribute to the weak ferromagnetic behavior.

Based on the magnetic hysteresis loops of the doped $BiFeO_3$ nanoparticles, the Pb doping in the $BiFeO_3$ structure seems to have a small effect on the magnetic properties. Substitution of Pb^{2+} ions at the Bi^{3+} sites induces oxygen vacancies which may lead to the enhancement of magnetic moments in the sample [83]. However,

Sample	Structure	Lattice parameters (Å)
$BiFeO_3$	Rhombohedral	a = b = 5.578 (1)
		c = 13.862 (3)
$Bi_{0.9}Pb_{0.1}FeO_3$	Cubic	a = b = c = 3.958 (1)
$BiFe_{0.9}Ni_{0.1}O_3$	Rhombohedral	a = b = 5.574 (1)
		c = 13.840 (4)

Table 1.
Rietveld analysis results for the XRD patterns of the undoped $BiFeO_3$ and doped $BiFeO_3$ ($Bi_{0.9}Pb_{0.1}FeO_3$ and $BiFe_{0.9}Ni_{0.1}O_3$) powders.

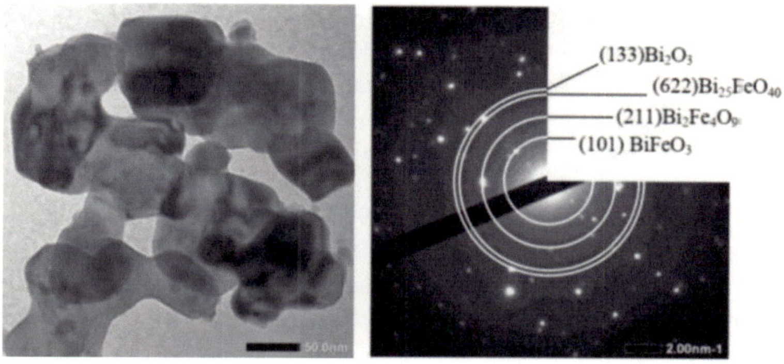

Figure 7.
TEM image with selected area electron diffraction (SAED) pattern for barium ferrite powders synthesized by sol-gel method and then calcined at 650°C for 1 h in air.

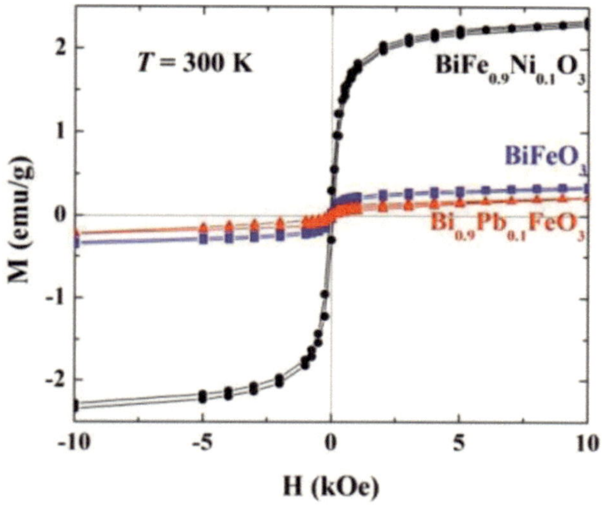

Figure 8.
Magnetic hysteresis curves of the undoped $BiFeO_3$ and doped $BiFeO_3$ ($Bi_{0.9}Pb_{0.1}FeO_3$ and $BiFe_{0.9}Ni_{0.1}O_3$) powders synthesized by the sol-gel method.

Verma and Kotnala [84] have confirmed through the SQUID measurements that $BiFeO_3$ with Pb doping exhibits a strong antiferromagnetism suggesting that the reduction of oxygen vacancies is realized in the system. Moreover, Ederer and Spaldin [85] have proposed that the magnetization value can be affected by the presence of oxygen vacancies but with a small change due to the formation of Fe^{2+} at the $BiFeO_3$ sites adjacent to the vacancy. Therefore, there is almost no increase in the magnetic parameters after Pb doping. Moreover, the weak ferromagnetism is commonly observed in the $Bi_{1-x}A_xFeO_3$ (A = Ca, Sr., Pb, Ba) system providing a canting of the antiferromagnetic sublattice [86], which is in line with this present work. On the other hand, Ni-doped $BiFeO_3$ nanoparticles show a significant increase on the magnetic parameters, namely, remanent and saturation magnetization This result is consistent with the previous paper by Hwang et al. [87], in which the Ni-doped $BiFeO_3$ sample exhibits similar rhombohedral perovskite structure compared to that of the undoped one and the magnetic properties show enhancement with respect to the undoped one. The increase in magnetic properties can occur due to the effect of nanoparticle surface area and ferromagnetic interaction exchange between neighboring Fe^{3+} and Ni^{3+} ions in the $BiFeO_3$ system [88].

The room temperature P-E loop of the prepared undoped bismuth ferrite, presented in **Figure 9**, exhibits unsaturated hysteresis loop. The curve was not fully saturated because of the low applied electric field. The remanent polarization (R_s) and the coercive field (E_c) of the undoped $BiFeO_3$ nanoparticles are about 20.5 μC/cm^2 and 5.5 V/cm, respectively. These values are lower than the values reported in the single crystal which has a large polarization of ~100 μC/cm^2 along (111) for bulk bismuth ferrite [89]. The existence of secondary phases, such as $Bi_{25}FeO_{40}$, $Bi_2Fe_4O_9$, and Bi_2O_3, affects the lower values of R_s and E_c in the sample. Pradhan et al. [78] have reported that leakage current is one of the major reasons for obtaining lower values of saturation polarization (P_s), R_s, and E_c in $BiFeO_3$ system.

In the Pb-doped $BiFeO_3$ nanoparticles, the Pb substitution improves the dielectric and ferroelectric properties [90]. It can be seen from **Table 2** that the electric properties, including dielectric constant, electrical conductivity, and electrical permittivity, increase with Pb doping in the $BiFeO_3$ crystal. It has been found that Pb substitution on the Bi site in the $BiFeO_3$ may destroy ferroelectricity ordering induced by Bi lone pair in the rhombic structure until it reaches a stable pseudocubic structure of $BiFeO_3$ [91]. In this work, addition of Pb doping in $BiFeO_3$ with x = 0.1 has already resulted in a pseudocubic structure, and, hence, the enhancement of the electrical properties is realized in the present sample. The value of dielectric constant with Pb doping, x = 0.1, at 1 kHz is in a good agreement with the work done by Zhang et al. [92]. The defect of oxygen vacancy due to Pb doping can increase the polarity of the sample and finally increase its dielectric constant. In addition, oxygen vacancy created as the consequence of Pb substitution on Bi site in the $BiFeO_3$ system plays an important role related to the ferroelectricity for Pb-doped $BiFeO_3$ sample. Moreover, the presence of Pb doping causes the existence of Fe^{2+} ion at Fe^{3+} sites which can produce holes around the Fe^{3+} site [93]. This effect is shown by the increasing value of electrical conductivity. It has been suggested that the relatively low number of oxygen vacancies in this sample may result in an improvement of the ferroelectric properties [94], as shown in **Table 2**.

As mentioned earlier, the Ni doping in $BiFeO_3$ nanoparticles enhances the magnetic properties as reported in the former paper [88]. However, the dielectric and

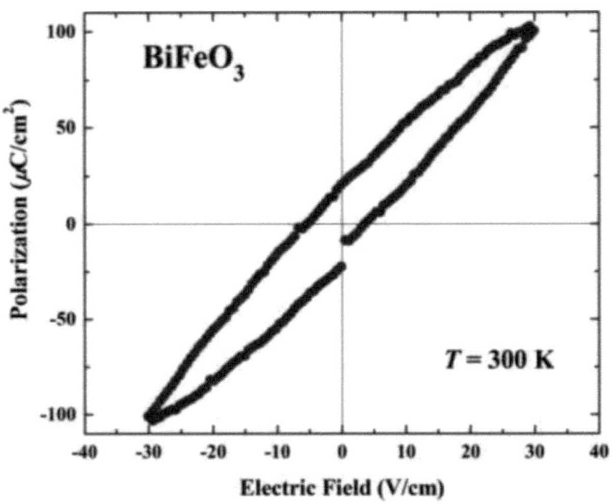

Figure 9.
Room temperature polarization-electric field (P-E) hysteresis loop of the undoped $BiFeO_3$ pellet sintered at 750°C.

Sample	Dielectric constant (ε_r) f = 1 kHz, T = 300 K	Conductivity [$\times 10^{-4}$ $(\Omega\,m)^{-1}$] T = 300 K	Permittivity ($\times 10^{-10}$ F/m) T = 300 K
$BiFeO_3$	19.4	0.012	1.7
$Bi_{0.9}Pb_{0.1}FeO_3$	130.8	0.162	11.6
$BiFe_{0.9}Ni_{0.1}O_3$	17.5	0.010	1.6

Table 2.
Dielectric constant, electrical conductivity, and permittivity of the undoped $BiFeO_3$ and doped $BiFeO_3$ ($Bi_{0.9}Pb_{0.1}FeO_3$ and $BiFe_{0.9}Ni_{0.1}O_3$) powders measured at room temperature.

other electrical properties of the Ni-doped $BiFeO_3$ have lower values than those of the undoped one, as displayed in **Table 2**. This means that the sample has inappropriate Ni doping concentration to improve the ferroelectricity. Moreover, the reduction in the dielectric constant is attributed to the decrease in the total polarization occurring in the sample. It is well known that the total polarization of a dielectric material is a combination of electronic, ionic, dipolar, and interfacial/space charge polarizations. The lower value of dielectric constant is probably caused by the effect of Ni doping on the ionic transformation from Fe^{2+} to be Fe^{3+} again. As the consequence of the charge stability, it may consume holes. Hence, the holes as charge carrier decrease. This is one reason of the decrease of sample's conductivity [95]. Another possible reason on decreasing value of electrical properties in Ni-doped $BiFeO_3$ sample is the impurity effect. It should be noticed that the impurity phases such as $Bi_2Fe_4O_9$ and $Bi_{25}FeO_{40}$ may also contribute to the electrical properties in $BiFeO_3$ [48]. The existence of multiphase in the sample leads to the increase of insulating grain boundaries affecting the electrical conductivity as well as the total polarization in the sample. The increase in the amount of grain boundaries, acting as the barrier for charge carrier mobility, results in the decrease of conductivity in the system.

8. Conclusions

Exploration related to the use of natural materials for functional materials has been applied in this study. Natural iron sand with the dominant magnetite (Fe_3O_4) content has been successfully synthesized through the chemical coprecipitation method as a starting material for producing hematite (α-Fe_2O_3). α-Fe_2O_3 has been successfully used as the source of Fe^{3+} ions to synthesize calcium ferrite and bismuth ferrite nanoparticles. The calcium ferrite powders synthesized by the chemical dissolved technique produce nano-sized crystals with the dominant phases of $CaFe_4O_7$ and $Ca_4Fe_{14}O_{25}$. The calcium ferrite powder has soft magnetic properties at room temperature which is attributed to the presence of dominant ferromagnetic phase and also oxygen vacancy in the nanoparticles. Magnetic parameters, such as saturation magnetic, are comparable to the barium-calcium hexaferrites, so that these nanoparticles have the potential application as microwave-absorbing materials. The bismuth ferrite powder, synthesized by the sol-gel method, exhibits multiferroic properties. The undoped $BiFeO_3$ possesses a weak ferromagnetism at room temperature. The magnetic parameters can be enhanced by Ni doping in the form of $BiFe_{0.9}Ni_{0.1}O_3$ nanoparticles. On the other hand, the electrical properties, i.e., dielectric constant, permittivity, and electrical conductivity, can be improved by Pb doping in the nanoparticles of $Bi_{0.9}Pb_{0.1}FeO_3$. The multiferroic behaviors are strongly determined by the nano-sized effects, the presence of oxygen vacancies and impurities, and also the doping type affecting the phase stability in the perovskite structure of $BiFeO_3$ crystals. Considering the importance of applying these ferrite-based nanoparticles, investigations for obtaining pure phases of the nanoparticles from natural resources are very important and need further study.

Acknowledgements

This work was partially supported by the grant of the International Research Collaboration, provided by DRPM, Ministry of Research, Technology and Higher Education, 2017–2019. We are thankful to Prof. Y. Kohori and Dr. H. Fukazawa, Chiba University, Japan, for the use of magnetic hysteresis measuring apparatus.

Conflict of interest

We state that the article is original and all authors are aware of its content and approve its submission. This article has not been published previously, and it is not under consideration for publication elsewhere. I confirm that there is no conflict of interest exists.

Author details

Malik Anjelh Baqiya[1], Retno Asih[1], Muhammad Ghufron[2], Mastuki[1], Dwi Yuli Retnowati[1], Triwikantoro[1] and Darminto[1*]

1 Department of Physics, Faculty of Science, Institut Teknologi Sepuluh Nopember, Surabaya, Indonesia

2 Department of Physics, Faculty of Mathematics and Natural Science, University of Brawijaya, Malang, Indonesia

*Address all correspondence to: darminto@physics.its.ac.id

IntechOpen

© 2019 The Author(s). Licensee IntechOpen. This chapter is distributed under the terms of the Creative Commons Attribution License (http://creativecommons.org/licenses/by/3.0), which permits unrestricted use, distribution, and reproduction in any medium, provided the original work is properly cited. (cc) BY

References

[1] Musyarofah, Lestari ND, Nurlaila R, Muwwaqor NF, Triwikantoro, Pratapa S. Synthesis of high-purity zircon, zirconia, and silica nanopowders from local zircon sand. Ceramics International. 2019;**45**(6):6639-6647. DOI: 10.1016/j.ceramint.2018.12.152

[2] Baqiya A, Munasir M, Zainuri M, Arifin Z, Mashuri M, Triwikantoro T, et al. Nanopowders produced from natural sources using the simple coprecipitation method. In: Bartul Z, Trenor J, editors. Advances in Nanotechnology. Vol. 20. New York: Nova Science Publisher; 2017. pp. 1-38

[3] Triwikantoro, Baqiya MA, Heriyanto T, Mashuri, Darminto D. Nano-coating of aluminum surface using Fe_3O_4-based magnetic fluids. Journal of Superconductivity and Novel Magnetism. 2017;**30**(2):555-560. DOI: 10.1007/s10948-016-3813-7

[4] Baqiya MA, Taufiq A, Sunaryono, Ayun K, Zainuri M, Pratapa S, et al. Spinel-structured nanoparticles for magnetic and mechanical applications. In: Magnetic Spinels-Synthesis, Properties and Applications. London: Intech Open; 2017. pp. 253-272

[5] Baqiya MA, Taufiq A, Sunaryono, Munaji, Sari PS, Dwihapsari Y, et al. Development of PVA/Fe_3O_4 as smart magnetic hydrogels for biomedical applications. In: Hydrogels. London: IntechOpen; 2018. pp. 159-178

[6] Dubey V, Kain V. Synthesis of magnetite by coprecipitation and sintering and its characterization. Materials and Manufacturing Processes. 2018;**33**(8):835-839. DOI: 10.1080/10426914.2017.1401720

[7] Lemine OM, Omri K, Zhang B, El Mir L, Sajieddine M, Alyamani A, et al. Sol-gel synthesis of 8nm magnetite (Fe_3O_4) nanoparticles and their magnetic properties. Superlattices and Microstructures. 2012;**52**(4):793-799. DOI: 10.1016/j.spmi.2012.07.009

[8] Lei W, Liu Y, Si X, Xu J, Du W, Yang J, et al. Synthesis and magnetic properties of octahedral Fe_3O_4 via a one-pot hydrothermal route. Physics Letters A. 2017;**381**(4):314-318. DOI: 10.1016/j.physleta.2016.09.018

[9] Wu JH, Ko SP, Liu HL, Jung M-H, Lee JH, Ju J-S, et al. Sub 5nm Fe_3O_4 nanocrystals via coprecipitation method. Colloids and Surfaces A: Physicochemical and Engineering Aspects. 2008;**313-314**:268-272. DOI: 10.1016/j.colsurfa.2007.04.108

[10] Lian S, Kang Z, Wang E, Jiang M, Hu C, Xu L. Convenient synthesis of single crystalline magnetic Fe_3O_4 nanorods. Solid State Communications. 2003;**127**(9):605-608. DOI: 10.1016/S0038-1098(03)00580-5

[11] Wei Y, Han B, Hu X, Lin Y, Wang X, Deng X. Synthesis of Fe_3O_4 nanoparticles and their magnetic properties. Procedia Engineering. 2012;**27**:632-637. DOI: 10.1016/j.proeng.2011.12.498

[12] Pan Y, Zeng W, Li L, Zhang Y, Dong Y, Ye K, et al. Surfactant assisted, one-step synthesis of Fe_3O_4 nanospheres and further modified Fe_3O_4/C with excellent lithium storage performance. Journal of Electroanalytical Chemistry. 2018;**810**:248-254. DOI: 10.1016/j.jelechem.2018.01.025

[13] Ramesh R, Rajalakshmi M, Muthamizhchelvan C, Ponnusamy S. Synthesis of Fe_3O_4 nanoflowers by one pot surfactant assisted hydrothermal method and its properties. Materials Letters. 2012;**70**:73-75. DOI: 10.1016/j.matlet.2011.11.085

[14] Arévalo P, Isasi J, Caballero AC, Marco JF, Martín-Hernández F. Magnetic and structural studies of Fe_3O_4 nanoparticles synthesized via coprecipitation and dispersed in different surfactants. Ceramics International. 2017;**43**(13):10333-10340. DOI: 10.1016/j.ceramint.2017.05.064

[15] Deepak FL, Bañobre-López M, Carbó-Argibay E, Cerqueira MF, Piñeiro-Redondo Y, Rivas J, et al. A systematic study of the structural and magnetic properties of Mn-, Co-, and Ni-doped colloidal magnetite nanoparticles. The Journal of Physical Chemistry C. 2015;**119**(21):11947-11957. DOI: 10.1021/acs.jpcc.5b01575

[16] Pereira C, Pereira AM, Fernandes C, Rocha M, Mendes R, Fernández-García MP, et al. Superparamagnetic MFe_2O_4 (M = Fe, Co, Mn) nanoparticles: Tuning the particle size and magnetic properties through a novel one-step coprecipitation route. Chemistry of Materials. 2012;**24**(8):1496-1504. DOI: 10.1021/cm300301c

[17] Saha P, Rakshit R, Mandal K. Enhanced magnetic properties of Zn doped Fe_3O_4 nano hollow spheres for better bio-medical applications. Journal of Magnetism and Magnetic Materials. 2019;**475**:130-136. DOI: 10.1016/j.jmmm.2018.11.061

[18] Kandasamy G, Maity D. Recent advances in superparamagnetic iron oxide nanoparticles (SPIONs) for in vitro and in vivo cancer nanotheranostics. International Journal of Pharmaceutics. 2015;**496**(2):191-218. DOI: 10.1016/j.ijpharm.2015.10.058

[19] Zhu M, Wang Y, Meng D, Qin X, Diao G. Hydrothermal synthesis of hematite nanoparticles and their electrochemical properties. The Journal of Physical Chemistry C. 2012;**116**(30):16276-16285. DOI: 10.1021/jp304041m

[20] Lassoued A, Dkhil B, Gadri A, Ammar S. Control of the shape and size of iron oxide (α-Fe_2O_3) nanoparticles synthesized through the chemical precipitation method. Results in Physics. 2017;**7**:3007-3015. DOI: 10.1016/j.rinp.2017.07.066

[21] Tadic M, Panjan M, Damnjanovic V, Milosevic I. Magnetic properties of hematite (α-Fe_2O_3) nanoparticles prepared by hydrothermal synthesis method. Applied Surface Science. 2014;**320**:183-187. DOI: 10.1016/j.apsusc.2014.08.193

[22] Liu X, Chen K, Shim J-J, Huang J. Facile synthesis of porous Fe_2O_3 nanorods and their photocatalytic properties. Journal of Saudi Chemical Society. 2015;**19**(5):479-484. DOI: 10.1016/j.jscs.2015.06.009

[23] Kefeni KK, Msagati TAM, Nkambule TTI, Mamba BB. Synthesis and application of hematite nanoparticles for acid mine drainage treatment. Journal of Environmental Chemical Engineering. 2018;**6**(2):1865-1874. DOI: 10.1016/j.jece.2018.02.037

[24] Lagoeiro LE. Transformation of magnetite to hematite and its influence on the dissolution of iron oxide minerals. Journal of Metamorphic Geology. 1998;**16**(3):415-423. DOI: 10.1111/j.1525-1314.1998.00144.x

[25] Umar Saeed K, Abdul M, Nasrullah K, Amir M, Abdur R, Amanullah. Transformation mechanism of magnetite nanoparticles. Materials Science-Poland. 2015;**33**(2):278-285. DOI: 10.1515/msp-2015-0037

[26] Ding C, Lv X, Li G, Bai C, Xuan S, Tang K, et al. Reaction routes

of $CaO–Fe_2O_3–TiO_2$ and calcium ferrite–TiO_2 system in continuous heating process. In: 9th International Symposium on High-Temperature Metallurgical Processing. Cham, Heidelberg: Springer International Publishing; 2018. pp. 159-165

[27] Phillips B, Muan A. Phase equilibria in the system CaO-iron oxide in air and at 1 atm. O_2 pressure. Journal of the American Ceramic Society. 1958;**41**(11):445-454. DOI: 10.1111/j.1151-2916.1958.tb12893.x

[28] Boyanov BS. Solid state interactions in the systems $CaO(CaCO_3)$-Fe_2O_3 and $CuFe_2O_4$-CaO. Journal of Mining and Metallurgy Section B: Metallugy. 2005;**41**(B1):67-77

[29] Saleh HI. Synthesis and formation mechanisms of calcium ferrite compounds. Journal of Materials Science and Technology. 2004;**20**(5):530-534

[30] Gill N, Puthucheri S. Bandwidth enhancement of calcium ferrite-graphite nanocomposite microwave absorber using single square loop frequency selective surface. In: 11th International Conference on Industrial and Information Systems (ICIIS). 2016. pp. 827-832

[31] Sulaiman NH, Ghazali MJ, Majlis BY, Yunas J, Razali M. Superparamagnetic calcium ferrite nanoparticles synthesized using a simple sol-gel method for targeted drug delivery. Biomedical Materials and Engineering. 2015;**26**(1):103-110

[32] Asenath-Smith E, Lokuhewa IN, Misture ST, Edwards DD. p-type thermoelectric properties of the oxygen-deficient perovskite $Ca_2Fe_2O_5$ in the brownmillerite structure. Journal of Solid State Chemistry. 2010;**183**(7):1670-1677. DOI: 10.1016/j.jssc.2010.05.016

[33] Dhankhar S, Bhalerao G, Ganesamoorthy S, Baskar K, Singh S. Growth and comparison of single crystals and polycrystalline brownmillerite $Ca_2Fe_2O_5$. Journal of Crystal Growth. 2017;**468**:311-315. DOI: 10.1016/j.jcrysgro.2016.09.051

[34] Phan TL, Tho PT, Tran N, Kim DH, Lee BW, Yang DS, et al. Crystalline and electronic structures and magnetic and electrical properties of La-doped $Ca_2Fe_2O_5$ compounds. Journal of Electronic Materials. 2018;**47**(1):188-195. DOI: 10.1007/s11664-017-5841-x

[35] Jijil CP, Lokanathan M, Chithiravel S, Nayak C, Bhattacharyya D, Jha SN, et al. Nitrogen doping in oxygen-deficient $Ca_2Fe_2O_5$: A strategy for efficient oxygen reduction oxide catalysts. ACS Applied Materials & Interfaces. 2016;**8**(50):34387-34395. DOI: 10.1021/acsami.6b11718

[36] Martin LW, Schlom DG. Advanced synthesis techniques and routes to new single-phase multiferroics. Current Opinion in Solid State and Materials Science. 2012;**16**(5):199-215. DOI: 10.1016/j.cossms.2012.03.001

[37] Safi R, Shokrollahi H. Physics, chemistry and synthesis methods of nanostructured bismuth ferrite ($BiFeO_3$) as a ferroelectro-magnetic material. Progress in Solid State Chemistry. 2012;**40**(1):6-15. DOI: 10.1016/j.progsolidstchem.2012.03.001

[38] Čebela M, Zagorac D, Batalović K, Radaković J, Stojadinović B, Spasojević V, et al. $BiFeO_3$ perovskites: A multidisciplinary approach to multiferroics. Ceramics International. 2017;**43**(1):1256-1264. DOI: 10.1016/j.ceramint.2016.10.074

[39] Wang X, Yang C, Zhou D, Wang Z, Jin M. Chemical co-precipitation synthesis and properties of pure-phase $BiFeO_3$. Chemical Physics Letters.

2018;**713**:185-188. DOI: 10.1016/j.cplett.2018.09.043

[40] Goldman AR, Fredricks JL, Estroff LA. Exploring reaction pathways in the hydrothermal growth of phase-pure bismuth ferrites. Journal of Crystal Growth. 2017;**468**:104-109. DOI: 10.1016/j.jcrysgro.2016.09.054

[41] Majid F, Mirza ST, Riaz S, Naseem S. Sol-gel synthesis of $BiFeO_3$ nanoparticles. Materials Today: Proceedings. 2015;**2**(10):5293-5297. DOI: 10.1016/j.matpr.2015.11.038

[42] Sankar Ganesh R, Sharma SK, Sankar S, Divyapriya B, Durgadevi E, Raji P, et al. Microstructure, structural, optical and piezoelectric properties of $BiFeO_3$ nanopowder synthesized from sol-gel. Current Applied Physics. 2017;**17**(3):409-416. DOI: 10.1016/j.cap.2016.12.008

[43] Wu H, Xue P, Lu Y, Zhu X. Microstructural, optical and magnetic characterizations of $BiFeO_3$ multiferroic nanoparticles synthesized via a sol-gel process. Journal of Alloys and Compounds. 2018;**731**:471-477. DOI: 10.1016/j.jallcom.2017.10.087

[44] Wang L, Xu J-B, Gao B, Bian L, Chen X-Y. Synthesis of pure phase $BiFeO_3$ powders by direct thermal decomposition of metal nitrates. Ceramics International. 2013;**39**:221-225. DOI: 10.1016/j.ceramint.2012.10.066

[45] Wang L, Xu J-B, Gao B, Chang A-M, Chen J, Bian L, et al. Synthesis of $BiFeO_3$ nanoparticles by a low-heating temperature solid-state precursor method. Materials Research Bulletin. 2013;**48**(2):383-388. DOI: 10.1016/j.materresbull.2012.10.038

[46] Zheng S, Wang J, Zhang J, Ge H, Chen Z, Gao Y. The structure and magnetic properties of pure single phase $BiFeO_3$ nanoparticles by microwave-assisted sol-gel method. Journal of Alloys and Compounds. 2018;**735**:945-949. DOI: 10.1016/j.jallcom.2017.10.133

[47] Majid F, Riaz S, Tariq N, Naseem S. Ferromagnetic behavior of undoped $BiFeO_3$ thin films prepared by sol-gel. Materials Today: Proceedings. 2015;**2**(10):5274-5279. DOI: 10.1016/j.matpr.2015.11.035

[48] Betancourt-Cantera LG, Bolarín-Miró AM, Cortés-Escobedo CA, Hernández-Cruz LE, Sánchez-De Jesús F. Structural transitions and multiferroic properties of high Ni-doped $BiFeO_3$. Journal of Magnetism and Magnetic Materials. 2018;**456**:381-389. DOI: 10.1016/j.jmmm.2018.02.065

[49] Chauhan S, Kumar M, Chhoker S, Katyal SC, Singh H, Jewariya M, et al. Multiferroic, magnetoelectric and optical properties of Mn doped $BiFeO_3$ nanoparticles. Solid State Communications. 2012;**152**(6):525-529. DOI: 10.1016/j.ssc.2011.12.037

[50] Costa LV, Deus RC, Foschini CR, Longo E, Cilense M, Simões AZ. Experimental evidence of enhanced ferroelectricity in Ca doped $BiFeO_3$. Materials Chemistry and Physics. 2014;**144**(3):476-483. DOI: 10.1016/j.matchemphys.2014.01.022

[51] Dhir G, Uniyal P, Verma NK. Multiferroic properties of Sr-doped $BiFeO_3$ nanoparticles. Physica B: Condensed Matter. 2018;**531**:51-57. DOI: 10.1016/j.physb.2017.12.004

[52] Tian Y, Xue F, Fu Q, Zhou L, Wang C, Gou H, et al. Structural and physical properties of Ti-doped $BiFeO_3$ nanoceramics. Ceramics International. 2018;**44**(4):4287-4291. DOI: 10.1016/j.ceramint.2017.12.013

[53] Wang Y, Guo Z, Jia Q, Dong J, Zhang J, Chen D. Effect of Nd/Mn

substitution on the structure and magnetic properties of nano-$BiFeO_3$. Journal of Alloys and Compounds. 2019;**786**:385-393. DOI: 10.1016/j.jallcom.2019.01.369

[54] Yuan X, Shi L, Zhao J, Zhou S, Li Y, Xie C, et al. Sr and Pb co-doping effect on the crystal structure, dielectric and magnetic properties of $BiFeO_3$ multiferroic compounds. Journal of Alloys and Compounds. 2017;**708**:93-98. DOI: 10.1016/j.jallcom.2017.02.288

[55] Asih R, Gufron M, Darminto. Synthesis of (Pb,Ni)-doped $BiFeO_3$ multiferroic systems via a sol-gel method and their magneto-electric properties. AIP Conference Proceedings. 2013;**1554**(1):50-53. DOI: 10.1063/1.4820281

[56] Asih R, Gufron M, Amrillah T, Arifani M, Hariyanto, Fitriyah N, et al. Effect of Pb doping on multiphase coexistence and magneto-electric properties of bismuth ferrite. AIP Conference Proceedings. 2014;**1617**(1):26-29. DOI: 10.1063/1.4897096

[57] Shokrollahi H. Magnetic, electrical and structural characterization of $BiFeO_3$ nanoparticles synthesized by co-precipitation. Powder Technology. 2013;**235**:953-958. DOI: 10.1016/j.powtec.2012.12.008

[58] Abushad M, Khan W, Naseem S, Husain S, Nadeem M, Ansari A. Influence of Mn doping on microstructure, optical, dielectric and magnetic properties of $BiFeO_3$ nanoceramics synthesized via sol-gel method. Ceramics International. 2019;**45**(6):7437-7445. DOI: 10.1016/j.ceramint.2019.01.035

[59] Costa LV, Rocha LS, Cortés JA, Ramirez MA, Longo E, Simões AZ. Enhancement of ferromagnetic and ferroelectric properties in calcium doped $BiFeO_3$ by chemical synthesis. Ceramics International. 2015;**41**(8):9265-9275. DOI: 10.1016/j.ceramint.2015.03.086

[60] Liu H, Pu Y, Shi X, Yuan Q. Dielectric and ferroelectric properties of $BiFeO_3$ ceramics sintered in different atmospheres. Ceramics International. 2013;**39**:S217-S220. DOI: 10.1016/j.ceramint.2012.10.065

[61] Arifin Z, Riyanto A, Lailiyah Q, Triwikantoro, Pratapa S, Darminto. Precipitated $CaCO_3$ with unique crystalline morphology prepared from limestone. Transactions of the Indian Ceramic Society. 2015;**74**(4):202-207. DOI: 10.1080/0371750X.2015.1084892

[62] Lutterotti L, Scardi P. Simultaneous structure and size-strain refinement by the Rietveld method. Journal of Applied Crystallography. 1990;**23**(4):246-252. DOI: 10.1107/S0021889890002382

[63] Rietveld H. A profile refinement method for nuclear and magnetic structures. Journal of Applied Crystallography. 1969;**2**(2):65-71. DOI: 10.1107/S0021889869006558

[64] Hughes H, Roos P, Goldring DC. X-ray data on some calcium-iron-oxygen compounds. Mineralogical Magazine and Journal of the Mineralogical Society. 2018;**36**(278):280-291. DOI: 10.1180/minmag.1967.036.278.10

[65] Tsuyuki N, Koizumi K, Umemura Y. Surface structure of converter slag stabilized by heating. Journal of the American Ceramic Society. 2007;**90**(1):225-229. DOI: 10.1111/j.1551-2916.2006.01344.x

[66] Ding C, Lv X, Chen Y, Bai C. Crystallization kinetics of $2CaO{\cdot}Fe_2O_3$ and $CaO{\cdot}Fe_2O_3$ in the $CaO–Fe_2O_3$ system. ISIJ International.

2016;**56**(7):1157-1163. DOI: 10.2355/isijinternational.ISIJINT-2015-710

[67] Millon E, Malaman B, Bonazebi A, Brice JF, Gerardin R, Evrard O. Structure cristalline du ferrite hemicalcique $CaFe_4O_7$. Materials Research Bulletin. 1986;**21**(8):985-994. DOI: 10.1016/0025-5408(86)90136-4

[68] Gerardin R, Millon E, Bonazebi A, Brice JF, Jeannot F, Evrard O. Structures et proprietes magnetiques des ferrites de calcium pseudo hexagonaux $Ca_4Fe_9O_{17}$, $CaFe_4O_7$ et $Ca_3Fe_{15}O_{25}$. Journal of Physics and Chemistry of Solids. 1988;**49**(4):343-348. DOI: 10.1016/0022-3697(88)90090-X

[69] Jotania RB, Khomane RB, Chauhan CC, Menon SK, Kulkarni BD. Synthesis and magnetic properties of barium–calcium hexaferrite particles prepared by sol-gel and microemulsion techniques. Journal of Magnetism and Magnetic Materials. 2008;**320**(6):1095-1101. DOI: 10.1016/j.jmmm.2007.10.032

[70] Samariya A, Dolia SN, Prasad AS, Sharma PK, Pareek SP, Dhawan MS, et al. Size dependent structural and magnetic behaviour of $CaFe_2O_4$. Current Applied Physics. 2013;**13**(5):830-835. DOI: 10.1016/j.cap.2012.12.009

[71] Ke H, Wang W, Wang Y, Xu J, Jia D, Lu Z, et al. Factors controlling pure-phase multiferroic $BiFeO_3$ powders synthesized by chemical co-precipitation. Journal of Alloys and Compounds. 2011;**509**(5):2192-2197. DOI: 10.1016/j.jallcom.2010.09.213

[72] Xu J-H, Ke H, Jia D-C, Wang W, Zhou Y. Low-temperature synthesis of $BiFeO_3$ nanopowders via a sol-gel method. Journal of Alloys and Compounds. 2009;**472**(1):473-477. DOI: 10.1016/j.jallcom.2008.04.090

[73] Carvalho TT, Tavares PB. Synthesis and thermodynamic stability of multiferroic $BiFeO_3$. Materials Letters. 2008;**62**(24):3984-3986. DOI: 10.1016/j.matlet.2008.05.051

[74] Sakar M, Balakumar S, Saravanan P, Jaisankar SN. Annealing temperature mediated physical properties of bismuth ferrite ($BiFeO_3$) nanostructures synthesized by a novel wet chemical method. Materials Research Bulletin. 2013;**48**(8):2878-2885. DOI: 10.1016/j.materresbull.2013.04.008

[75] Yang N, Yang H, Jia J, Pang X. Formation and magnetic properties of nanosized $PbFe_{12}O_{19}$ particles synthesized by citrate precursor technique. Journal of Alloys and Compounds. 2007;**438**(1):263-267. DOI: 10.1016/j.jallcom.2006.08.037

[76] Zhu XH, Defaÿ E, Lee Y, André B, Aïd M, Zhu JL, et al. High permittivity $Bi_{24}Fe_2O_{39}$ thin films prepared by a low temperature process. Applied Physics Letters. 2010;**97**(23):232903. DOI: 10.1063/1.3524492

[77] Sun L, Zhang R, Wang Z, Ju L, Cao E, Zhang Y. Structural, dielectric and magnetic properties of $NiFe_2O_4$ prepared via sol-gel auto-combustion method. Journal of Magnetism and Magnetic Materials. 2017;**421**:65-70. DOI: 10.1016/j.jmmm.2016.08.003

[78] Pradhan AK, Zhang K, Hunter D, Dadson JB, Loiutts GB, Bhattacharya P, et al. Magnetic and electrical properties of single-phase multiferroic $BiFeO_3$. Journal of Applied Physics. 2005;**97**(9):093903. DOI: 10.1063/1.1881775

[79] Jia D-C, Xu J-H, Ke H, Wang W, Zhou Y. Structure and multiferroic properties of $BiFeO_3$ powders. Journal of the European Ceramic Society. 2009;**29**(14):3099-3103. DOI: 10.1016/j.jeurceramsoc.2009.04.023

[80] Ederer C, Spaldin NA. Weak ferromagnetism and magnetoelectric coupling in bismuth ferrite. Physical Review B. 2005;**71**(6):060401. DOI: 10.1103/PhysRevB.71.060401

[81] Jaffari GH, Aftab M, Samad A, Mumtaz F, Awan MS, Shah SI. Effects of dopant induced defects on structural, multiferroic and optical properties of $Bi_{1-x}Pb_xFeO_3$ ($0 \leq x \leq 0.3$) ceramics. Materials Research Express. 2018;**5**(1):016103. DOI: 10.1088/2053-1591/aaa14f

[82] Makhdoom AR, Akhtar MJ, Rafiq MA, Siddique M, Iqbal M, Hasan MM. Enhancement in the multiferroic properties of $BiFeO_3$ by charge compensated aliovalent substitution of Ba and Nb. AIP Advances. 2014;**4**(3):037113. DOI: 10.1063/1.4869081

[83] Khomchenko VA, Kiselev DA, Kopcewicz M, Maglione M, Shvartsman VV, Borisov P, et al. Doping strategies for increased performance in $BiFeO_3$. Journal of Magnetism and Magnetic Materials. 2009;**321**(11):1692-1698. DOI: 10.1016/j.jmmm.2009.02.008

[84] Verma KC, Kotnala RK. Tailoring the multiferroic behavior in $BiFeO_3$ nanostructures by Pb doping. RSC Advances. 2016;**6**(62):57727-57738. DOI: 10.1039/C6RA12949H

[85] Ederer C, Spaldin NA. Influence of strain and oxygen vacancies on the magnetoelectric properties of multiferroic bismuth ferrite. Physical Review B. 2005;**71**(22):224103. DOI: 10.1103/PhysRevB.71.224103

[86] Khomchenko VA, Kiselev DA, Selezneva EK, Vieira JM, Lopes AML, Pogorelov YG, et al. Weak ferromagnetism in diamagnetically-doped $Bi_{1-x}A_xFeO_3$ (A = Ca, Sr, Pb, Ba) multiferroics. Materials Letters. 2008;**62**(12):1927-1929. DOI: 10.1016/j.matlet.2007.10.044

[87] Hwang JS, Yoo YJ, Lee YP, Kang J-H, Lee KH, Lee BW, et al. Reinforced magnetic properties of Ni-doped $BiFeO_3$ ceramic. Journal of the Korean Physical Society. 2016;**69**(3):282-285. DOI: 10.3938/jkps.69.282

[88] Zhao J, Zhang X, Liu S, Zhang W, Liu Z. Effect of Ni substitution on the crystal structure and magnetic properties of $BiFeO_3$. Journal of Alloys and Compounds. 2013;**557**:120-123. DOI: 10.1016/j.jallcom.2013.01.005

[89] Ederer C, Spaldin NA. Recent progress in first-principles studies of magnetoelectric multiferroics. Current Opinion in Solid State and Materials Science. 2005;**9**(3):128-139. DOI: 10.1016/j.cossms.2006.03.001

[90] Mazumder R, Sen A. Effect of Pb-doping on dielectric properties of $BiFeO_3$ ceramics. Journal of Alloys and Compounds. 2009;**475**(1):577-580. DOI: 10.1016/j.jallcom.2008.07.082

[91] Chaigneau J, Haumont R, Kiat JM. Ferroelectric order stability in the $Bi_{1-x}Pb_xFeO_3$ solid solution. Physical Review B. 2009;**80**(18):184107. DOI: 10.1103/PhysRevB.80.184107

[92] Zhang X, Sui Y, Wang X, Tang J, Su W. Influence of diamagnetic Pb doping on the crystal structure and multiferroic properties of the $BiFeO_3$ perovskite. Journal of Applied Physics. 2009;**105**(7):07D918. DOI: 10.1063/1.3079770

[93] Kianinia M, Ahadi K, Nemati A. Investigation of dark and light conductivities in calcium doped bismuth ferrite thin films. Materials Letters. 2011;**65**(19):3086-3088. DOI: 10.1016/j.matlet.2011.06.052

[94] Kumar A, Sharma P, Yang W, Shen J, Varshney D, Li Q. Effect of La and Ni substitution on structure, dielectric and ferroelectric properties of $BiFeO_3$ ceramics. Ceramics International. 2016;**42**(13):14805-14812. DOI: 10.1016/j.ceramint.2016.06.113

[95] Nadeem M, Khan W, Khan S, Husain S, Ansari A. Tailoring dielectric properties and multiferroic behavior of nanocrystalline $BiFeO_3$ via Ni doping. Journal of Applied Physics. 2018;**124**(16):164105. DOI: 10.1063/1.5050946

Chapter 4

Surface Plasmons in Oxide Semiconductor Nanoparticles: Effect of Size and Carrier Density

Hiroaki Matsui

Abstract

Oxide semiconductors have received much attention for potential use in optoelectronic applications such as transparent electrodes, transistors, and emitting devices. Recently, new functionalities of oxide semiconductors have been discovered such as localized surface plasmon resonances (LSPRs), which show high-efficiency plasmon excitations in the infrared (IR) range using different structures such as nanorods, nanoparticles (NPs), and nanodots. In this chapter, we introduce optical properties of carrier- and size-dependent LSPRs in oxide semiconductor NPs based on In_2O_3: Sn (ITO). In particular, systematic examinations of carrier- and size-dependent LSPRs reveal the damping mechanisms on LSPR excitations of ITO NPs, which play an important role in determining excitation efficiency of LSPRs. Additionally, the control of carrier and size in the ITO NPs contribute toward improving solar-thermal shielding in the IR range. The high IR reflectance of assembled films of ITO NPs is due to three-dimensional plasmon coupling between the NPs, which is related to electron carriers and particle size of ITO NPs. This chapter provides new information concerning structural design when fabricating thermal-shielding materials based on LSPRs in oxide semiconductor NPs.

Keywords: oxide semiconductor, nanocrystal, plasmon, infrared, energy-saving

1. Introduction

Metals (such as Au and Ag) have been utilized for the majority of plasmonic materials in the visible range. Recently, oxide semiconductors have attracted much attention for use as potential new plasmonic materials. In particular, ZnO: Ga and In_2O_3: Sn (ITO) are known for use as transparent electrodes due to their metallic conductivity. These oxide semiconductors show surface plasmon resonances (SPRs) in the infrared (IR) range [1, 2]. Propagated SPRs can be excited on metal surfaces using a prism-coupling technique such as a Kretschmann-type attenuated total reflection (ATR) system [3]. Our research group has investigated the optical properties of SPRs excited on ZnO: Ga and ITO film surfaces from the viewpoint of physical characteristics such as field strength and penetration depth [4–6]. On the other hand, subwavelength materials such as nanorods, nanoparticles (NPs), and nanodots are capable of supporting localized surface plasmon resonances (LSPRs), which can be directly excited by incident light in the absence of a prism-coupling method [7, 8]. Above all, LSPRs confined to NPs can lead to light at the nanoscale

 IntechOpen

when confining the collective oscillations of free electrons into NPs. This LSPR effect further provides strong electric fields (*E*-fields) on NP surfaces, which contribute to surface-enhanced optical spectroscopy [9]. For example, assembled films consisting of ITO NPs have demonstrated optical enhancements of near-IR luminescence and absorption in the IR range [10, 11]. Therefore, optical studies concerning oxide semiconductor NPs can break new research ground in the area of plasmonics and metamaterials.

An understanding of plasmon damping is very important in order to achieve high-efficiency LSPRs. A number of plasmonic studies of metal NPs have been devoted to investigating the damping processes of LSPRs. For metal NPs, there are two main damping processes, comprising (i) size-dependent surface scattering and (ii) electronic structure-related inter- and intraband damping [12–15]. The damping processes are closely related to the physical properties of the metals. Therefore, understanding of the damping processes of LSPRs in oxide semiconductor NPs is also important for the control of optical properties. Oxide semiconductor NPs are useful plasmonic materials since their LSPR wavelengths can be widely tuned by electron density in addition to particle size [16–18]. Carrier control of LSPRs indicates that oxide semiconductors have an additional means of tuning the optical properties in a manner that is not as readily available for metal NPs. In particular, carrier-dependent damping is a specific feature of the plasmonic response in oxide semiconductor NPs. Precise elucidation of the carrier-dependent damping process including structural size is required for the optical design of plasmonic materials based on oxide semiconductor NPs.

The purpose of this chapter is to report on the light interactions of size- and carrier-controlled ITO NPs and to discuss their plasmonic applications in the IR range. We introduce size- and carrier-dependent plasmonic responses and provide information for the physical interpretation of optical spectra. A rigorous approach to the analysis of the optical properties allows us to show a quantitative assessment of the electronic properties in ITO NPs. The employments of Mie theoretical calculations, which can describe well the optical properties of metal NPs, are validated in terms of ITO NPs. Finally, we discuss the optical properties assembled films of ITO NPs for solar-thermal shielding.

2. Carriers and plasmon excitations

2.1 Synthesis of ITO NPs

ITO NPs with different Sn contents were fabricated using the chemical thermolysis method with various initial ratios of precursor complexes $(C_{10}H_{22}O_2)_3In$ and $(C_{10}H_{22}O_2)_4Sn$ [19]. Indium and tin complexes were thermal heated at 300–350°C for 4 h in a reducing agent, and the mixture was then gradually cooled to room temperature. The resultant mixture produced a pale blue suspension and to which was then added excess ethanol to induce precipitation. Centrifugation and repeated washing were conducted four times using ethanol, which produced dried powders of ITO NPs with a pale blue color. Finally, the powder samples were dispersed in a nonpolar solvent of toluene. Electrophoresis analysis revealed a positive zeta potential of +31 meV for the NPs, which indicated the NPs had non-aggregated states in the solvent due to electrostatic repulsion between NPs. Particle surfaces of the NPs were terminated by organic ligands consisting of fatty acids, which contributed in spatial separation between NPs.

2.2 Carrier-dependent plasmon absorptions

Optical absorptions and TEM images of ITO NPs with different electron densities (n_e) were examined (**Figure 1**). TEM images revealed that all NP sizes (D) were ca. 36 nm (**Figure 2(a–c)**). This indicates that the systematic change in the absorption spectra is related to the Sn content. Absorption measurements were performed using a Fourier-transform infrared (FT-IR) spectrometer. A value of n_e was estimated from the absorption spectra by theoretical calculations. The following equation was used to derive absorption intensity (A) from the experimental data [20]:

$$A = 4\pi k R^3 \mathrm{Im}\left\{\frac{\varepsilon_m(\omega) - \varepsilon_d}{\varepsilon_m(\omega) + 2\varepsilon_d}\right\} \quad (1)$$

where $k = 2\pi(\varepsilon_d)^{1/2}\omega/c$ with c representing the speed of light, ε_d indicates the host dielectric constants of toluene, $\varepsilon_m(\omega)$ is the particle dielectric function, and R is

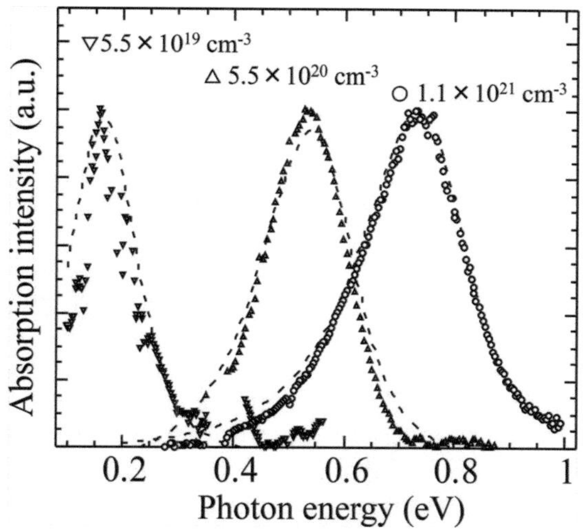

Figure 1.
Absorption spectra of ITO NPs with different electron densities. Doping with Sn contents of 0.02, 1, and 5% into the NPs' induced electron density of 6.3×10^{19} cm^{-3}, 5.7×10^{20} cm^{-3}, and 1.1×10^{21} cm^{-3}, respectively. Dot lines indicate theoretical calculations based on the modified Mie theory [19].

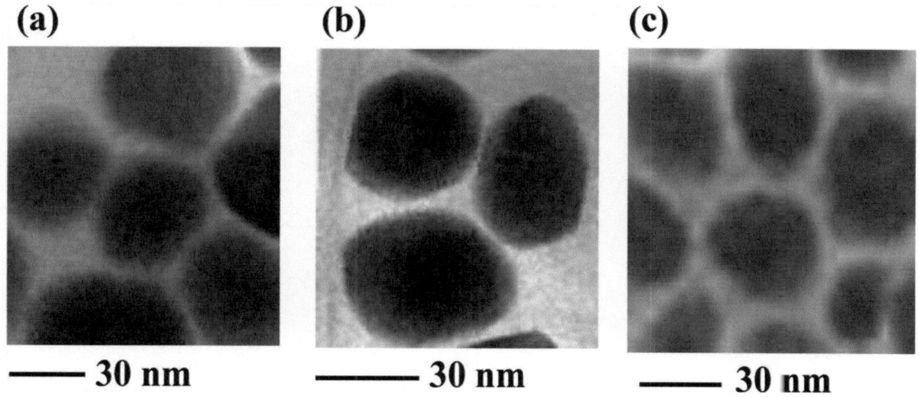

Figure 2.
TEM images of ITO NPs with electron densities of (a) 6.3×10^{19} cm^{-3}, (b) 5.7×10^{20} cm^{-3}, and (c) 1.1×10^{21} cm^{-3} [19].

the particle radius. Furthermore, $\varepsilon_m(\omega)$ employed the free-electron Drude term with frequency-dependent damping constant, $\Gamma(\omega)$, on the basis that ITO comprised free-electron carriers [20]:

$$\varepsilon_m(\omega) = 1 - \frac{\omega_p^2}{\omega(\omega + i\Gamma)} \tag{2}$$

The plasma frequency (ω_p) is given by $\omega_p^2 = ne/\varepsilon_\infty\varepsilon_0 m^*$, where ε_∞ is the high-frequency dielectric constant, ε_0 is the vacuum permittivity, and m^* is the effective electron mass. Fitted absorptions were used with parameter values of ε_d = 2.03 (n = 1.426 refractive index of the solvent), ε_∞ = 3.8, and m^* = 0.3 m_0 to estimate $\varepsilon_p(\omega)$.The term $\Gamma(\omega)$ based on electron-impurity scattering can be described by the following relation [21]:

$$\Gamma(\omega) = f(\omega)\Gamma_L + [1 - f(\omega)]\Gamma_H\left(\frac{\omega}{\Gamma_H}\right)^{-3/2} \tag{3}$$

where $f(\omega)$ can be described by $f(\omega)$ = $[1 + \exp\{(\omega - \Gamma_x)/\sigma\}]^{-1}$. Γ_H and Γ_L represent the high-frequency ($\omega = \infty$) and low-frequency (ω = 0) damping, respectively. Γ_X and σ represent the change-over frequency and width of the function, respectively.

Calculated absorption spectra were very close to the experimental data. ITO NPs doped with Sn content of 0.02, 1, or 5% provided electron density of 6.3×10^{19}, 5.7×10^{20}, and 1.1×20^{21} cm^{-3}, respectively (**Figure 1**). We summarized the LSPR resonant peak and absorption intensity as a function of n_e (**Figure 3(a)**). The LSPR resonant peak gradually showed a redshift from the near-IR to mid-IR range with decreasing n_e. Additionally, the absorption intensity decreased markedly with decreasing n_e. No plasmon excitation was observed in the low n_e region below 10^{19} cm^{-3}. The Mott critical density (N_c) of ITO is estimated as $N_c = 6 \times 10^{18}$ cm^{-3} (**Figure 3(b)**). Below the Mott critical density, the impurity band is not overlapped with the Fermi energy (E_F) level. ITO results in a band insulator.

However, the E_F level combined with the impurity band in the middle n_e region from 10^{19} to 10^{20} cm^{-3}. At the high n_e region above 10^{20} cm^{-3}, the E_F level is placed in a highest occupied state in the conduction band (CB). As a consequence, ITO

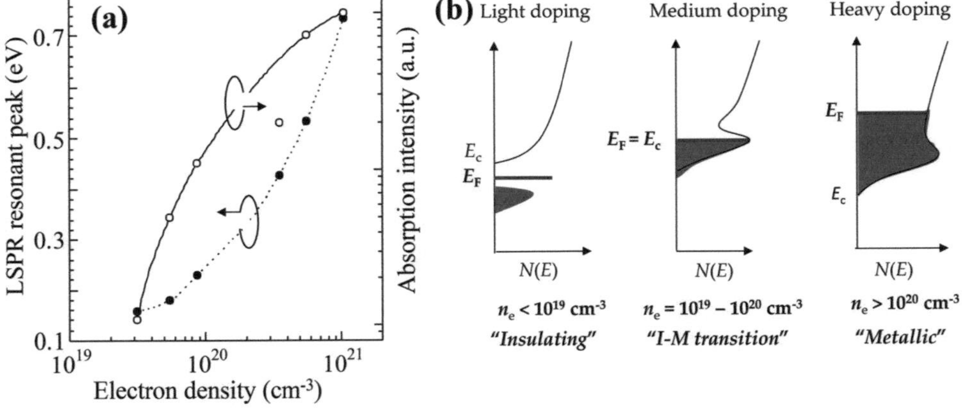

Figure 3.
(a) LSPR resonant peak and absorption intensity of ITO NPs as a function of electron density. (b) a schematic picture of electronic structures of ITO with different ranges of electron density.

shows metallic behavior. These results indicated that a large amount of free electrons were required to excite highly efficient plasmon excitations. ITO NPs were suitable for plasmonic materials in the near-IR range.

2.3 Damping mechanism

The two types of damping processes that exist in plasmon excitations of metal NPs are (i) bulk damping and (ii) surface damping. Bulk damping (γ_B) is related to electron-electron ($\gamma_{e\text{-}e}$), electron-phonon ($\gamma_{e\text{-}ph.}$), and electron-impurity scattering ($\gamma_{e\text{-impurity}}$). These scattering components determine a mean free path (l_m) of a free electron. On the other hand, surface scattering is effective when a NP size is smaller than l_m, which becomes the main damping process in NPs.

Surface scattering (γ_s) can be described by $\gamma_s = Av_F/l_{SC}$ for a small nanoparticle, where A is a material constant and v_F is the Fermi velocity [$v_F = \hbar/m^*(3\pi n_e)^{1/3}$]. The surface scattering length (l_{SC}) is defined by $l_{SC} = 4\ V/S$, where V is the volume and S is the surface area of the particle [20]. For our ITO NPs, l_{SC} was calculated as 24 nm, which was longer than the l_m of ITO (~10 nm) [22, 23]. For ITO NPs, no surface scattering was effective because the l_m of ITO was smaller than l_{SC}. Therefore, it is considered that ITO NPs are mainly related to bulk damping.

Metallic conductivity of ITO NPs is obtained by doping with impurity atoms, suggesting that ITO NPs involve electron-impurity scattering in bulk damping. The spectral features of ITO NPs could be fitted using Mie theory with frequency-dependent damping parameter $\Gamma(\omega)$. **Figure 4(a)** shows absorption spectra of ITO NPs with lowest (5.5×10^{19} cm^{-3}) and highest (1.1×10^{21} cm^{-3}) n_e values. For NPs with the lowest n_e, a symmetric absorption spectrum was obtained, while an asymmetric spectrum was obtained for NPs with the highest n_e. These spectral features were determined by Γ_H and Γ_L. **Figure 4(b)** shows the dependence of Γ_H and Γ_L on electron density. A difference in Γ_H and Γ_L values was found in the high n_e region above 10^{20} cm^{-3}. Electron-impurity scattering is reflected by Γ_L, providing asymmetric LSPR features by broadening in the low photon energy regions. In contrast, the Γ_L values (~70 meV) were the same as those of Γ_H in the low n_e region below 10^{20} cm^{-3}, indicating that LSPRs were independent of electron-impurity scattering.

The carrier-dependent plasmon response is divided into two n_e regions. Region-I comprises low n_e below 10^{20} cm^{-3}, in which coherence of electron oscillation in ITO NPs is not always disturbed by electron-impurity scattering. The spectral features of LSPRs comprise narrow line-widths and symmetric line-shapes. However, absorption intensity is small (**Figure 3(a)**) since a short mean free path length (l_m = 3–4 nm) determines the coherence of electron oscillations in the NPs. This situation is due to insufficient conduction paths. Region-II comprises high n_e above 10^{20} cm^{-3}, in which LSPR excitations become more effective with increasing l_m, as a result of increased n_e. The l_m value of NPs with the highest n_e was estimated as 10.7 nm. However, LSPR excitations are influenced by electron-impurity scattering, which generated the asymmetric line-shapes.

Degenerated metals on doped oxide semiconductors are generally realized by extrinsic and/or intrinsic dopants. However, the carrier screening effect from background cations is weak in contrast to metals with a short screening length (comprising several angstroms) [24]. Electron-impurity scattering dominates the optical properties of LSPRs in the high n_e region. In this work, the maximum lm in ITO NPs was 10.7 nm. Previous reports have detailed long l_m values from 14 to 16 nm on ITO films [22, 23]. Control of crystallinity and impurities in ITO NPs will be required to obtain high-efficiency LSPR excitations in the IR range.

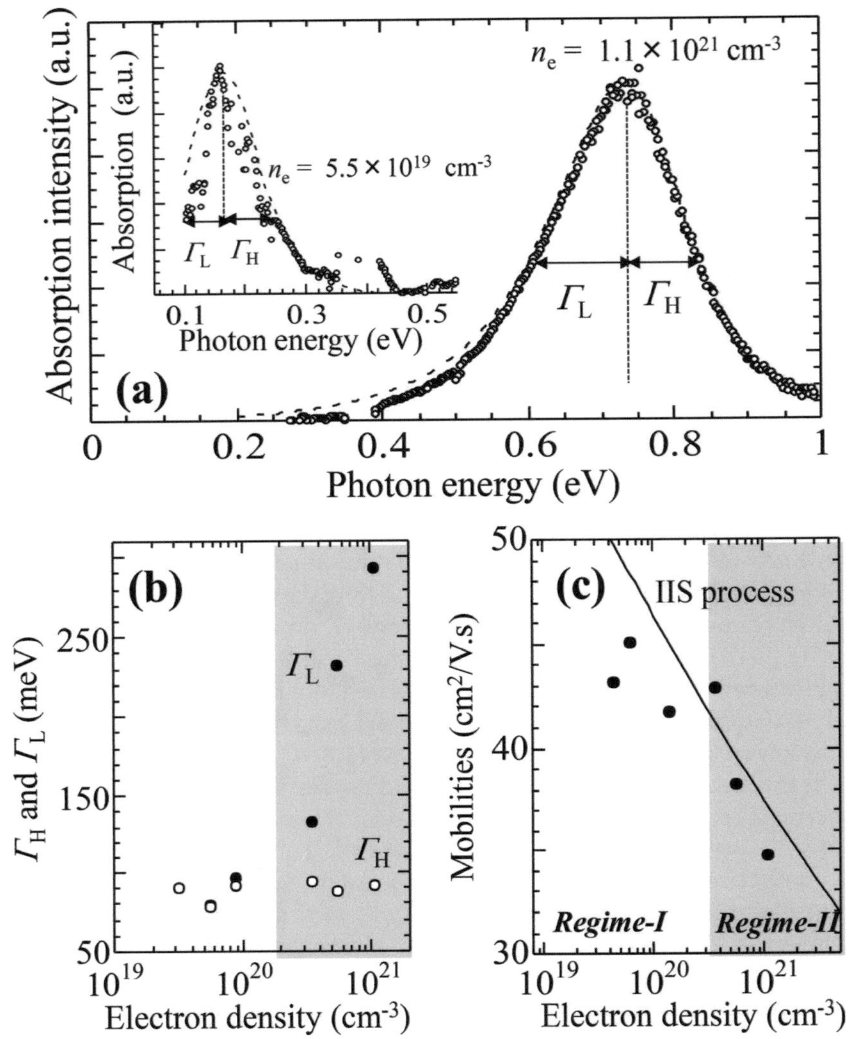

Figure 4.
(a) Absorption spectra of ITO NPs with n_e values of 5.5×10^{19} and 1.1×10^{21} cm^{-3}. (b) Dependence of Γ_H (•) and Γ_L (○) on electron density. (c) Mobility (μ_e) as a function of electron density. The μ_e (black dots) are compared with those obtained using ionized impurity scattering (IIS) process (black line).

3. Particle size and plasmon excitations

Figure 5(a) shows the size distribution of ITO NPs, revealing that size distribution gradually increased with increasing particle size (D): D = 10 ± 2.2 nm, 20 ± 3.5 nm, and 36 ± 4.3 nm. **Figure 6** shows TEM results of the dependency of NPs on particle size. In particular, NPs with D = 36 nm showed well-developed facet surfaces, and NPs were clearly separated from one another due to the presence of organic ligands formed on the NP surfaces. All NP samples showed broad peak characteristic of colloid NPs with a crystalline nature (**Figure 5(b)**). Patterns were similar to those of standard cubic bixbyite, which had no discernible SnO or SnO_2 peak. Besides, the line-width of the (222) peak, $\Delta(2\theta)$, was narrower for the NPs with D = 36 nm than D = 10 nm. These results reflected differences in crystallinity, size, defects, and strain in the NPs.

The absorption spectra of the NPs with different particle sizes are shown in **Figure 7(a)**. Based on the Mie theory with frequency-dependent damping, the

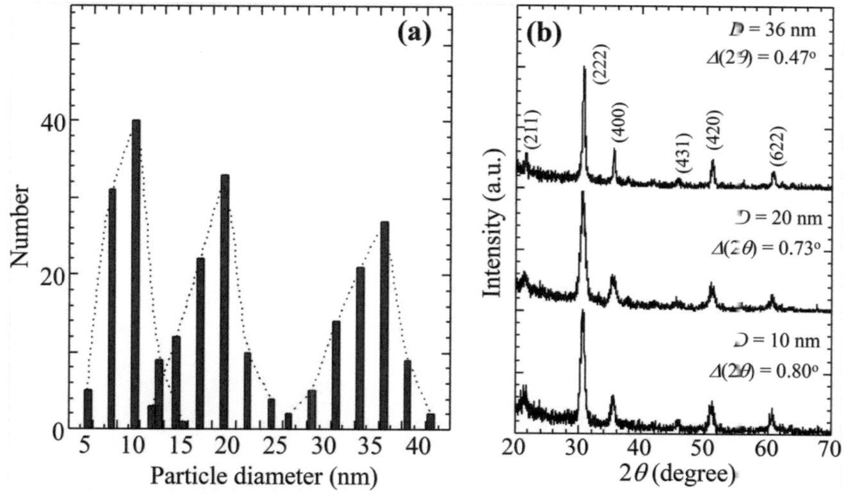

Figure 5.
(a) Size distributions of ITO NPs with particle sizes (D) of 10, 20, and 36 nm. Inset images show TEM images of ITO NPs with different particle sizes. (b) XRD 2q-q pattern of ITO NPs with D = 10, 20, and 36 nm. Δ(2θ) indicates a line-width of the (222) peak [25].

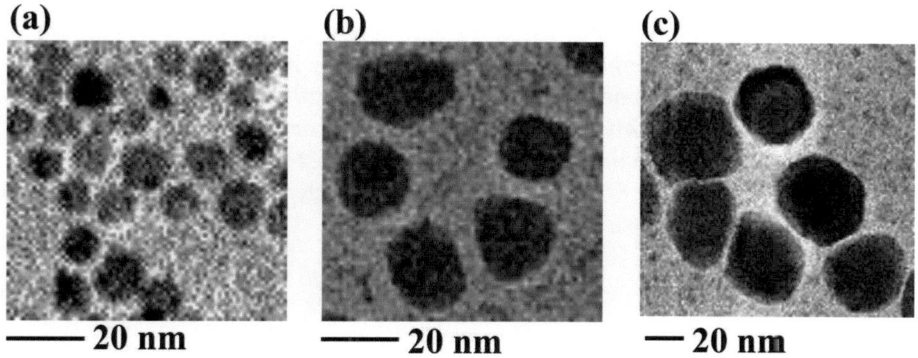

Figure 6.
TEM images of ITO NPs with D = 10 nm (a), 20 nm (b), and 36 nm (c) [25].

values of n_e were approximately 10^{21} cm^{-3}, and μ_e ranged from 21 to 37 $cm^2/V.s$. The broadening of the absorption spectra was related to the quality factor (Q-factor) of the plasmonic resonance defined by the ratio of peak energy to spectral linewidth of the LSPR peak. This factor provided a good indication of weak electronic damping and efficient *E*-field generation. *Q*-factor values of LSPRs with *D* = 10, 20, and 36 nm NPs were 2.4, 3.3, and 4.5 respectively. The increase in particle size is expected for strong *E*-field enhancement on the NP surfaces. It was indicated that the *Q*-factor values in the LSPR peaks were attributed to the electronic and crystalline properties. On the other hand, the LSPR peak positions were independent of particle size.

The peak positions of LPRs generally depend on the particle size in the case of metal NPs. The size-dependent absorption spectra of spherical NPs can be calculated precisely using the full Mie equations. These equations can describe well the size effects of LSPRs in metal NPs as follows.

An analytical solution to Maxwell's equations describes the extinction and scattering of light by spherical particles. The electromagnetic field produced by a plane wave incident on a homogeneous conducting sphere can be expressed by the following relations [26]:

$$\sigma_{ext} = \frac{2\rho}{|k|^2}\sum_{L=1}^{\infty}(2L+1)[\mathrm{Re}(a_L + b_L)] \tag{4}$$

where k is the incoming wave vector and L are integers representing the dipole, quadrupole, and higher multipoles of the scattering. In the above equations, a_L and b_L are represented by the following parameters, composed of the Riccati-Bessel functions ψ_L and χ_L [26]:

$$b_L = \frac{\psi_L(mx)\psi_L'(x) - m\psi_L'(mx)\psi_L(x)}{\psi_L(mx)\chi_L'(x) - m\psi_L'(mx)\chi_L(x)} \tag{5}$$

$$a_L = \frac{m\psi_L(mx)\psi_L'(x) - \psi_L'(mx)\psi_L(x)}{m\psi_L(mx)\chi_L'(x) - \psi_L'(mx)\chi_L(x)} \tag{6}$$

Here, $m = \tilde{n}/n_m$, where $\tilde{n} = n_R + in_I$ is the complex refractive index of the metal and n_m is the refractive index of the surrounding medium. Additionally, $x = k_m r$, where r is the radius of the particle. It should be noted that $k_m = 2\pi/\lambda_m$ is defined as the wavenumber in the medium rather than the vacuum wavenumber. Peak positions of absorption spectra of ITO NPs were estimated using the full Mie theory (black line in **Figure 7(b)**). The dielectric constants were taken from the ellipsometric data of an ITO film with an electron density of 1.0×10^{21} cm^{-3}. The estimated peak positions remained almost unchanged with particle sizes below 120 nm and then slightly redshifted to longer wavelengths with particle sizes above 120 nm. That is, ITO NPs with particle sizes below 40 nm had no high-order plasmon mode and were mainly dominated by light absorptions. These results differed largely from those of metal NPs. LSPR properties of ITO NPs could be fully described using Mie theory in the quasi-static limit.

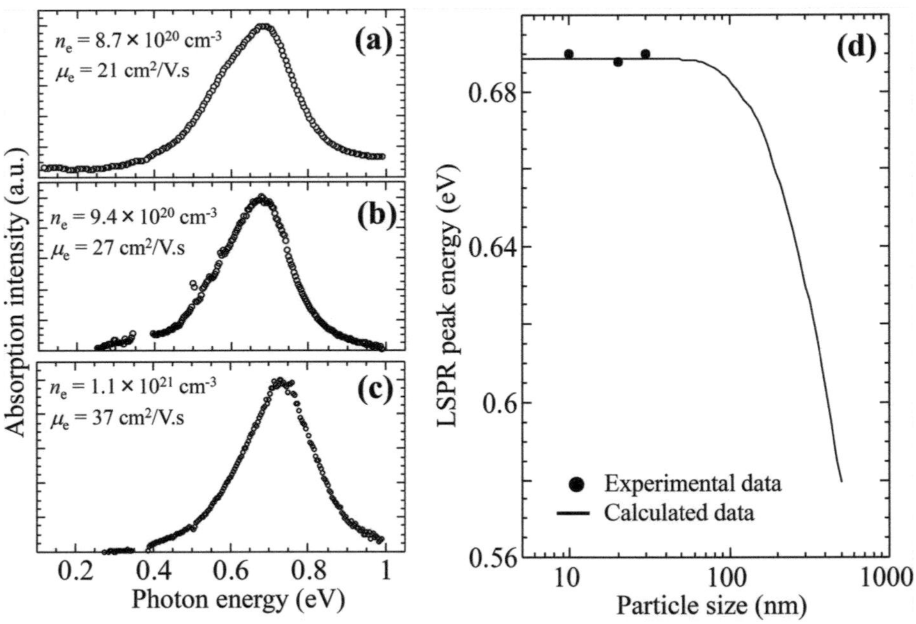

Figure 7.
(a) Absorption spectra of ITO NPs with different sizes comprising (a) 10 nm, (b) 20 nm, and (c) 36 nm. (d) LSPR peak energy as a function of particle size. A black line represents using Eqs. (4)–(6) [25].

4. Infrared applications for solar-thermal shielding

4.1 High reflections in the IR range

Recently, plasmonic properties on oxide semiconductors have attracted much attention in the area of solar-thermal shielding. The purpose of our study is to apply the plasmonic properties of assembled films of ITO NPs. To date, IR optical responses have been investigated with regard to transmittance and extinction spectra of composites and films using oxide semiconductor NPs. IR shielding properties by transmittance and absorption properties have mainly been discussed [27–30]. Reports concerning reflective performances in assemblies of NPs have yet to appear in spite of the desire for thermal shielding to cut IR radiation, not by absorption, but through reflection properties.

Assemblies of Ag and Au NPs can produce high *E*-fields through plasmon coupling between NPs in the visible range and are utilized in surface-enhanced spectroscopy [31, 32]. The high *E*-fields localized between NPs are very sensitive to interparticle gaps [33]. A gap length down to distances less than the size of a NP causes remarkable enhancements in *E*-fields. Surfactant- and additive-treated NPs are effective strategies that can be employed to obtain small interparticle gaps between NPs, which can be developed into one-, two-, and three-dimensional assemblies of NPs [34]. In particular, optical applications based on NPs have the benefit of large-area fabrications with lower costs to make NP assemblies attractive for industrial development.

In this section, we report on the plasmonic properties of assembled films comprising ITO NPs (ITO NP films) and their solar-thermal applications in the IR range [35]. Both experimental and theoretical approaches were employed in an effort to understand the plasmonic properties of the NP films. The IR reflectance of the NP films was analyzed on the basis of variations in particle size and electron density. The investigation focused in particular on *E*-field interactions in order to determine how the NP films affected high IR reflectance. This behavior is discussed in terms of the physical concept of plasmonic hybridization, which further clarified the importance of interparticle gaps for high IR reflectance.

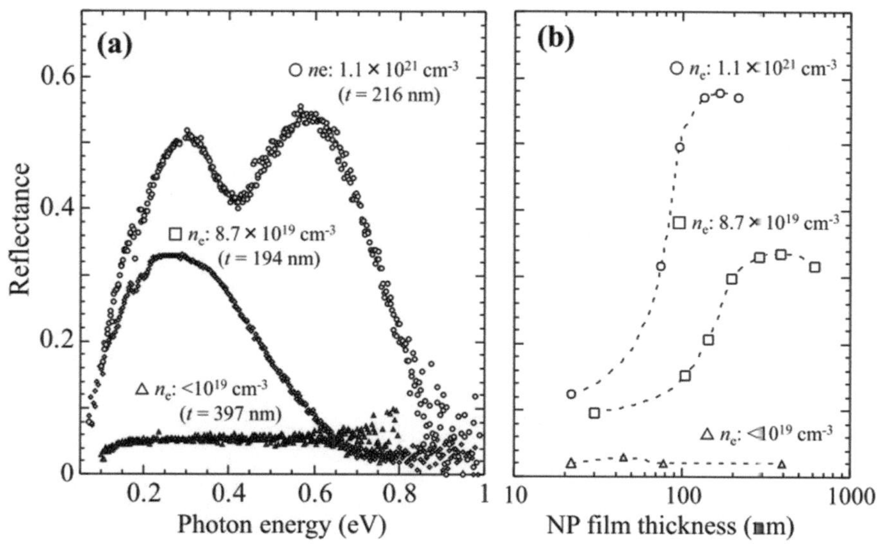

Figure 8.
(a) Reflectance spectra of ITO NP films with different electron densities of 1.1×10^{21} cm^{-3} *(○),* 8.7×10^{19} cm^{-3} *(□), and* $< 10^{19}$ cm^{-3} *(Δ). (b) Reflectance as a function of NP film thickness of ITO NP films with different electron densities.*

Figure 8(a) shows reflectance spectra of ITO NP films with different electron densities. The assembled ITO NP films were deposited on IR-transparent CaF_2 substrates through a spin-coating technique. The spin-coating conditions comprised sequential centrifugation at (i) 800 rpm for 5 s, (ii) 2400 rpm for 30 s, and (iii) 800 rpm for 10 s. The fabricated NP films were then thermally treated at 150°C in air to evaporate the solvent. Reflectance was enhanced with increasing electron density and reached a value of ca. 0.6 in the NP film with $n_e = 1.1 \times 10^{21}$ cm^{-3}. Additionally, reflectance was dependent on film thickness (**Figure 8(b)**). Reflectance gradually increased with increasing film thickness and was then saturated in film thicknesses above 200 nm. As a result, it is necessary to use NPs with high electron density in order to obtain NP films with high IR reflectance.

Figure 9(a) shows reflectance spectra of ITO NP films with different particle sizes. Reflectance gradually increased with increasing particle size, which was dependent on NP film thickness (**Figure 9(b)**). That is, increasing in particle size contributed to obtain high IR reflectance. Highly efficient solar-thermal shielding played an important role in controlling electron density and particle size. We found that the high IR reflectance was closely related to plasmon coupling between the NPs in the NP films as follows.

4.2 Electric field distributions

Figure 10(a) shows experimental and theoretical absorption spectra of ITO NPs dispersed in toluene. The theoretical data was simulated using the finite-difference time-domain (FDTD) method and was close to the experimental data. We observed the formation of a strong electric field (E-field) on the NP surface (inset of **Figure 10(a)**). The relationship between the *E*-field and photon energy was further investigated, as shown in **Figure 10(b–d)**. The *E*-field on the NP surface increased with increasing photon energy. A high E-field was obtained at an LSPR peak position of 1.8 μm. The LSPRs of ITO NPs produced the strong E-field on the NP surface.

We evaluated the optical properties of ITO NP films from the viewpoint of electrodynamic simulations based on the finite-difference time-domain (FDTD) method (**Figure 11(a)**). The modeled NP layer was assumed to have a hexagonally close-packed (HCP) structure with an interparticle distance (*r*) of 2 nm along the in-plane

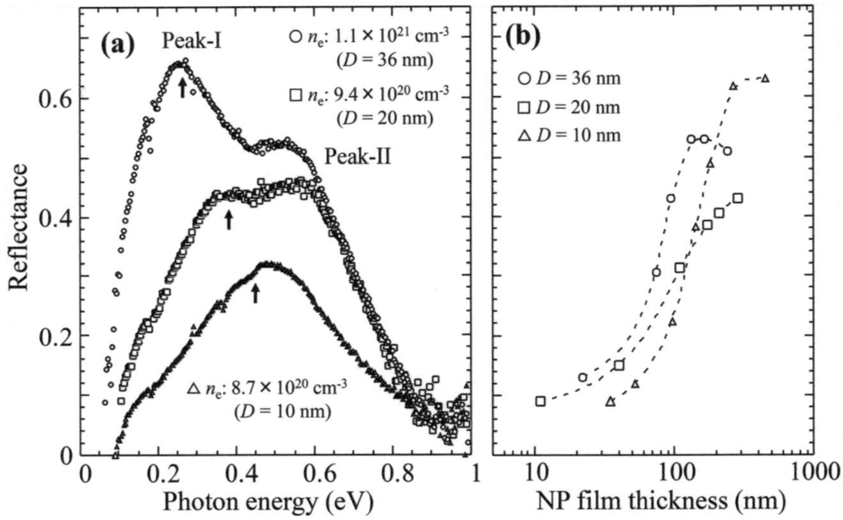

Figure 9.
(a) Reflectance spectra of ITO NP films with different particle sizes of 36 nm (○), 20 nm (□), and 10 nm (△). (b) Reflectance as a function of NP film thickness of ITO NP films with different particle sizes.

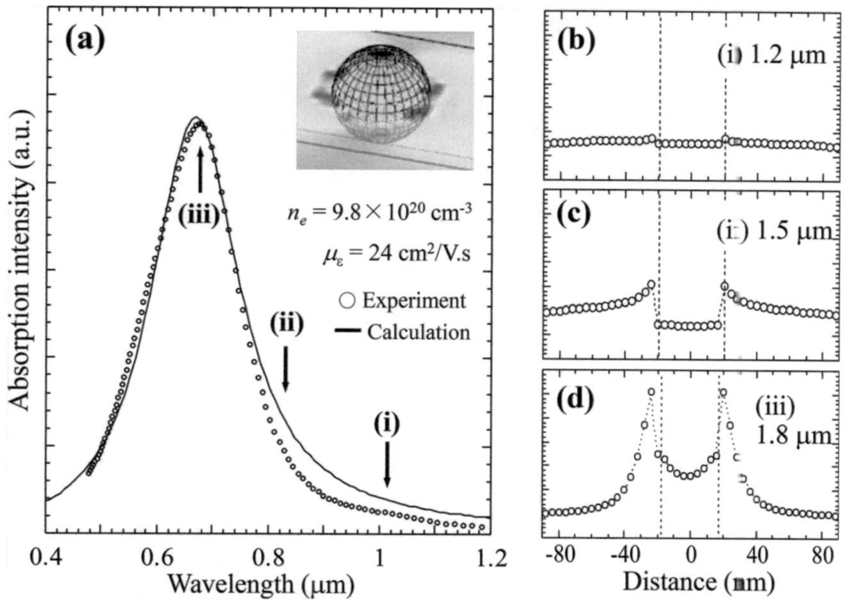

Figure 10.
(a) Absorption spectra of ITO NPs: Experimental (open circles) and simulated data (solid line). Inset indicates an electric field distribution on the NP surface obtained by the FDTD simulation. Cross-section field distributions at 1.2 μm (b), 1.5 μm (c), and 1.8 μm of the NPs.

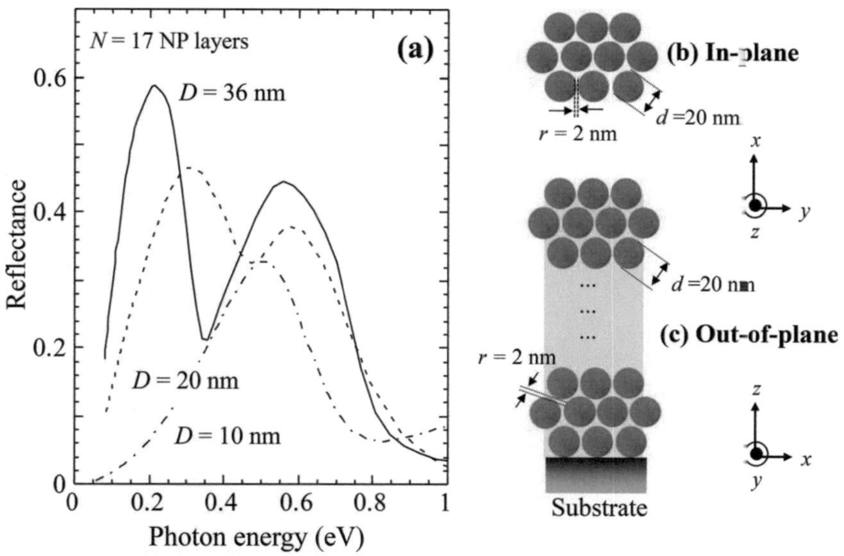

Figure 11.
(a) Simulated reflectance spectra of ITO NP layers at different particle sizes (D). A number of NP layer (N) was set to N = 20 NP layers. (b) and (c) indicate structural diagrams of a simulated NP layers along the in-plane (x-y) and out-of-plane (x-z) directions, respectively. The modeled structure was assumed to have a HCP structure with an interparticle distance (r) of 2 nm and was illuminated with light directed in the z direction from the air side. The E-field was parallel to the x direction.

(*x*-*y*) and out-of-plane (*y*-*z*) directions (**Figure 11(b and c)**). The modeled sample was illuminated with light directed in the *z* direction from the air side. The E-field was parallel to the x direction. The refractive index (real part: 1.437) of capric acid was used for the medium between the NPs. The dielectric functions of the ITO NPs were obtained from the parameter fitting for the absorption spectra. **Figure 11(a)** shows the reflectance spectra of ITO NP layers with different particle sizes (*D*) of 10, 20, and

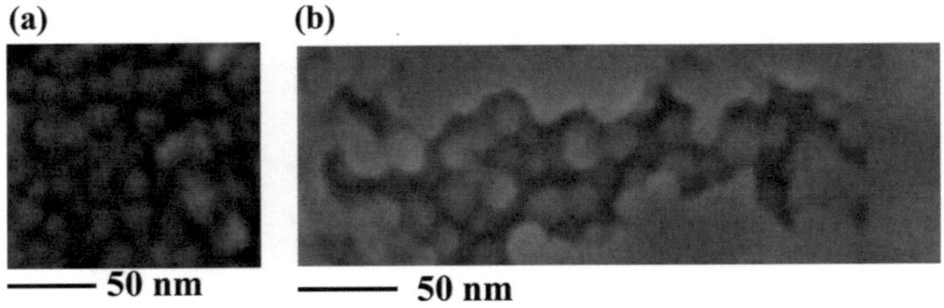

Figure 12.
SEM images of an ITO NP film along the in-plane (a) and out-of-plane (b) directions.

36 nm. The number of the NP layer was set to the $N = 20$ NP layer. Reflectance clearly enhanced with increasing particle size, which appeared as a result of three-dimensional assemblies of ITO NPs, and it was suggested theoretically that increasing particle size contributed to the reflective-type thermal shielding in the IR range.

Plasmon coupling between NPs produces large enhancements of E-fields at interparticle gaps. We typically investigated the E-field distributions at peak-II (0.60 eV) and peak-I (0.208 eV) for a 20 NP layer with $D = 36$ nm. **Figure 12(a and b)** shows SEM images of ITO NP films ($D = 36$ nm) along the in-plane and out-of-plane directions, revealing that the NPs had close-packed structures along both

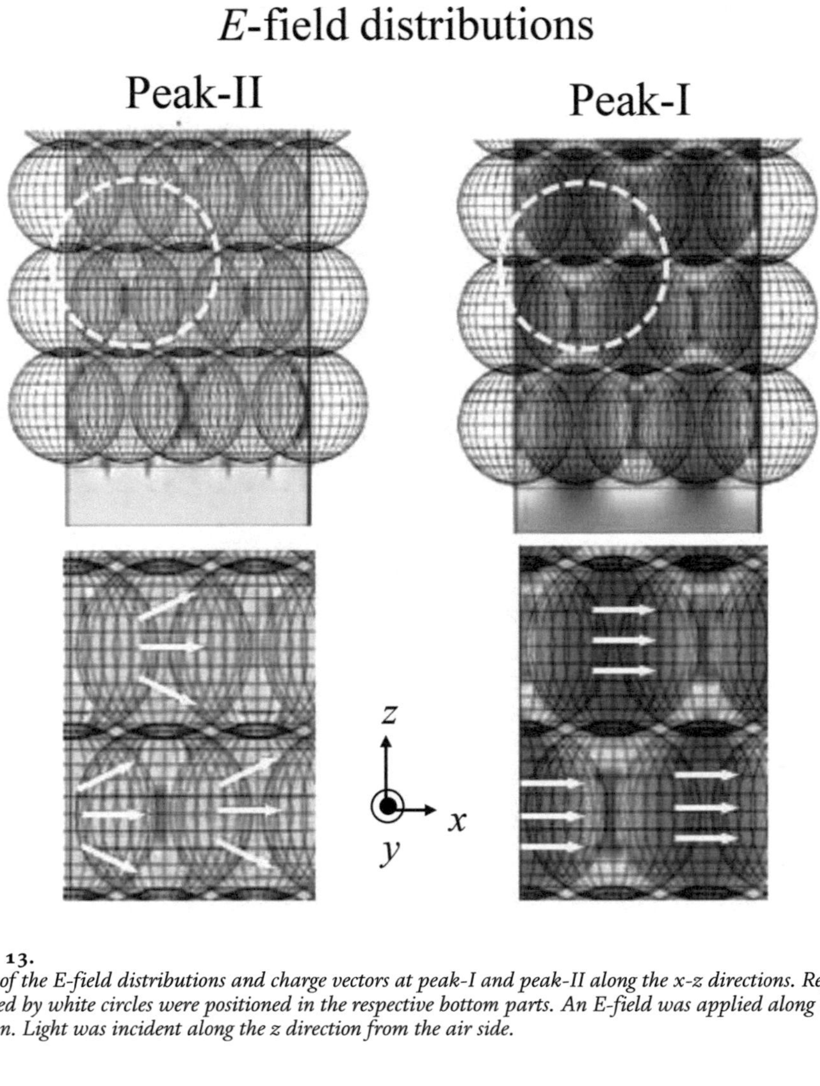

Figure 13.
Images of the E-field distributions and charge vectors at peak-I and peak-II along the x-z directions. Regions delimited by white circles were positioned in the respective bottom parts. An E-field was applied along the x direction. Light was incident along the z direction from the air side.

directions. **Figure 13** shows the E-field distributions along the x-z directions. For peak-I, the E-field between the NPs was strongly localized along the x direction when an electric field of light was applied along this direction. In contrast, peak-II displays E-fields along the diagonal directions in the x-z plane in addition to those along the x direction. A difference in the E-field of peak-I and peak-II was clearly found. The FDTD simulations revealed that the two types of reflectance peaks had different mechanisms of plasmon excitations. Therefore, it was indicated that different E-field distributions between the NPs played an important role in producing the IR reflectance in the IR range.

5. Conclusion

Optical properties of carrier- and size-dependent LSPRs were investigated using dopant-controlled ITO NPs. From systematic correlations between LSPR excitations and electron density, plasmon damping of ITO NPs was closely related to electron-impurity scattering, which was effective with high n_e values greater than 10^{20} cm^{-3}. That is, the role of electron carriers in ITO NPs could enhance LSPRs with simultaneous damped plasmon excitations. Changes in particle size also affected the LSPRs in ITO NPs. Increasing particle size altered the magnitude and peak splitting of the resonant reflectance, which covered a wide IR range. As a result, the carrier and size control of ITO NPs led to high solar-thermal shielding. The origin of the high IR reflectance of ITO NP films was clarified by electrodynamic simulations (FDTD). We found that the E-field distributions between the NPs along the in-plane and out-of-plane directions played key roles in producing the high IR reflectance. Control of electron carrier and particle size revealed important aspects that should be considered in the area of structural design when fabricating thermal-shielding materials.

Acknowledgements

This research was supported in part by a grant from JST A-Step (No. VP30218088667) and for Grant-in-Aids for Scientific Research (B) (No. 18H01468).

Conflict of interest

The authors declare no competing financial interest.

Author details

Hiroaki Matsui
Department of Bioengineering, Department of Electrical Engineering and Information Systems, The University of Tokyo, Tokyo, Japan

*Address all correspondence to: hiroaki@ee.t.u-tokyo.ac.jp

IntechOpen

© 2019 The Author(s). Licensee IntechOpen. This chapter is distributed under the terms of the Creative Commons Attribution License (http://creativecommons.org/licenses/by/3.0), which permits unrestricted use, distribution, and reproduction in any medium, provided the original work is properly cited.

References

[1] Dastmalchi B, Tassin P, Koschny T, Soukoulis M. A new perspective on plasmonics: Confinement and propagation length of surface plasmons for different materials and geometries. Advanced Optical Materials. 2015;**4**(1):1-8. DOI: 10.1002/adom.201500446

[2] Khamh H, Sachet E, Kelly K, Maria JP, Franzen S. As good as gold and better: Conducting metal oxide materials for mid-infrared plasmonic applications. Journal of Materials Chemistry C. 2018;**6**:8326-8342. DOI: 10.1039/c7tc05760a

[3] Pgeifer P, Adinger U, Schwotzer G, Diekmann S, Steinrücke P. Real time sensing of specific molecular binding using surface plasmon resonance spectroscopy. Sensors and Actuators B. 1999;**54**:166-175. DOI: 10.1016/S0925-4005(98)00334-7

[4] Matsui H, Badalawa W, Ikehata A, Tabata H. Oxide surface plasmon resonance for a new sensing platform in the near-infrared range. Advanced Optical Materials. 2013;**1**:397-403. DOI: 10.1002/adom.201200075

[5] Matsui H, Ikehata A, Tabata H. Surface plasmon sensors on ZnO: Ga layer surfaces: Electric field distributions and absorption-sensitive enhancements. Applied Physics Letters. 2015;**106**:019905. DOI: 10.1063/1.4905211

[6] Matsui H, Ikehata A, Tabata H. Asymmetric plasmon structures on ZnO: Ga for high sensitivity in the infrared range. Applied Physics Letters. 2016;**109**:191601. DOI: 10.1063/1.4966598

[7] Lounis SD, Runnerstrom EL, Llordés A, Milliron DJ. Defect chemistry and plasmon physics of colloidal metal oxide nanocrystals. Journal of Physical Chemistry Letters. 2014;**5**:1564-1574. DOI: 10.1021/jz500440e

[8] Li SQ, Guo P, Buchholz DB, Zhou W, Hua Y, Odom TW, et al. Plasmonic-photonic mode coupling in indium-tin-oxide nanorods arrays. ACS Photonics. 2014;**1**:163-172. DOI: 10.1021/ph400038g

[9] Agrawal A, Kriegel I, Milliron DJ. Shape-dependent field enhancement and plasmon resonance of oxide nanocrystals. Journal of Physical Chemistry C. 2015;**119**:6227-6238. DOI: 10.1021/acs.jpcc.5b01648

[10] Furube A, Yoshinaga T, Kanehara M, Eguchi M, Teranishi T. Electric field enhancement inducing near-infrared two photon absorption in an indium tin oxide nanoparticle film. Angewandte Chemie. 2012;**51**:2640-2642. DOI: 10.1002/anie.201107450

[11] Matsui H, Badalawa W, Hasebe T, Furuta S, Nomura W, Yatsui T, et al. Coupling of Er light emissions to plasmon modes on In_2O_3: Sn nanoparticle sheets in the near-infrared range. Applied Physics Letters. 2104;**105**:041903. DOI: 10.1063/1.4892004

[12] Sönnichsen C, Franzl T, Wilk T, Plessen GV, Feldmann J. Drastic reduction of plasmon damping in gold nanorods. Physical Review Letters. 2002;**88**:077402. DOI: 10.1103/PhysRevLett.88.077402

[13] Sönnichsen C, Franzl T, Plessen GV, Feldmann J. Plasmon resonances in large noble-metal clusters. New Journal of Physics. 2002;**4**:93.1-93.8. DOI: 10.1088/1367-2630/4/1/393

[14] Berciand S, Cognet L, Tamarat P, Lounis B. Observation of intrinsic size effects in the optical response of individual gold nanoparticles. Nano Letters. 2005;**5**:515-518. DOI: 10.1021/nl050062t

[15] Langhammer C, Schwind M, Kasemo B, Zorić. Localized surface plasmon resonances in aluminum nanodisks. Nano Letters. 2008;**8**: 1461-1471. DOI: 10.1021/nl080453i

[16] Runnerstrom EL, Bergerud A, Agrawal A, Johns RW, Dahlman CJ, Singh A, et al. Defect engineering in plasmonic metal oxide nanocrystals. Nano Letters. 2016;**16**:3390-3398. DOI: 10.1021/acs.nanolett6b01171

[17] Rowe DJ, Jeong JS, Mkhoyan KA, Kortshagen UR. Phosphorus-doped silicon nanocrystals exhibiting mid-infrared localized surface plasmon resonance. Nano Letters. 2013;**13**: 1317-1332. DOI: 10.1021/nl4001184

[18] Schimpf AM, Lounis SD, Runnerstrom EL, Milliron DJ, Gamelin DR. Redox chemistries and plasmon energies of Photodoped In_2O_3 and Sn-doped In_2O_3 (ITO) nanocrystals. Journal of the American Chemical Society. 2015; **137**:518-524. DOI: 10.1021/ja5116953

[19] Matsui H, Furuta S, Tabata H. Role of electron carriers on local surface plasmon resonances in doped oxide semiconductor nanocrystals. Applied Physics Letters. 2014;**104**:211903. DOI: 10.1063/1.4880356

[20] Hartland GV. Optical studies of dynamics in noble metal nanostructures. Chemical Reviews. 2011;**111**:3857-3887. DOI: 10.1021/cr1002547

[21] Pisarkiewicz Y, Zakrzewska K, Leja E. Scattering of charge carriers in transport and conducting thin oxide films with a non-parabolic conduction band. Thin Solid Films. 1989;**174**: 217-223. DOI: 10.1016/0040-6090(89) 90892-4

[22] Pammi SV, Chanda A, Ahn JK, Park JH, Cho CR, Lee CR, et al. Low resistivity ITO thin films deposited by NCD technique at low temperature: Variation of tin concentration. Journal of the Electrochemistry of Society. 2010;**157**:H937-H941. DOI: 10.1149/ 1.3467802

[23] Mohammed HA. Effect of substrate temperature on physical properties of In_2O_3: Sn films deposited by e-beam technique. International Journal of Physical Sciences. 2012;**7**:20120-22109. DOI: 10.5897.IJPS12.182.

[24] Ku HY, Ullman FG. Capacitance of thin dielectric structures. Journal of Applied Physics. 1964;**35**:265. DOI: 10.1063/1.1713297

[25] Matsui H, Hasebe T, Hasuike N, Tabata H. Plasmonic heat shielding in the infrared range using oxide semiconductor nanoparticles based on Sn-doped In_2O_3: Effect of size and interparticle gap. ACS Applied Nano Materials. 2018;**1**:1853-1862. DOI: 10.1021/acsanm8b00260

[26] Mie G. Beiträge zur optik trüber medien, speziell kolloidaler metallösungen. Ann. Phys. 1908;**330**: 377. DOI: 10.1002/amdp19083300302

[27] Li SY, Niklasson GA, Granqvist CG. Plasmon-induced near-infrared electrochromism based on transparent conducting nanoparticles: Approximate performance limits. Applied Physics Letters. 2012;**101**:071903. DOI: 10.1063/ 1.4739792

[28] Katagiri K, Takabatake R, Inumaru K. Robust infrared-shielding coating films prepared using perhydropolysilazane and hydrophobized indium tin oxide nanoparticles with tuned surface plasmon resonances. ACS Applied Materials & Interfaces. 2013;**5**:10240-10245. DOI: 10.1021/am403011t

[29] Fang X, Mak CL, Dai J, Li K, Ye H, Leung CW. ITO/Au/ITO sandwich structure for near-infrared plasmonics. ACS Applied Materials & Interfaces. 2014;**6**:15743-15752. DOI: 10.1021/ am50261565

[30] Tao P, Viswanath A, Schadler LS, Benicewicz BC, Siegel RW. Preparation and properties of indium tin oxide/epoxy nanocomposites with Polyglycidyl methacrylate grafted nanoparticles. ACS Applied Materials & Interfaces. 2011;**3**:3638-3645. DOI: 10.1021/am200841n

[31] Tao AR, Ceperley DP, Sinsermsuksakul P, Neureuther AR, Yang P. Self-organized silver nanoparticles for three-dimensional plasmonic crystals. Nano Letters. 2008; **8**:4033-4038. DOI: 10.1021/nl802877h

[32] Chen CF, Tzeng SD, Chen HY, Lin KJ, Gwo S. Tunable plasmonic response from alkanethiolate-stabilized gold nanoparticle superlattices: Evidence of near-field coupling. Journal of the American Chemical Society. 2008;**130**: 824-826. DOI: 10.1021/ja0773610

[33] Le F, Brandl DW, Urzhumov YA, Wang H, Lundu J, Halas NJ, et al. Metallic nanoparticle arrays: A common substrate for both surface-enhanced Raman scattering and surface-enhanced infrared absorption. ACS Nano. 2008;**2**: 707-718. DOI: 10.1021/nn800047e

[34] Toma M, Toma K, Michioka K, Ikezoe Y, Obara D, Okamoto K, et al. Collective plasmon modes excited on a silver nanoparticle 2D crystalline sheet. Physical Chemistry Chemical Physics. 2011;**13**:7459-7466. DOI: 10.1039/cocp02953j

[35] Matsui H, Furuta S, Hasebe T, Tabata H. Plasmonic-field interactions at nanoparticle interfaces for infrared thermal-shielding applications based on transparent oxide semiconductors. ACS Applied Materials & Interfaces. 2016;**8**: 11749-11757. DOI: 10.1021/acsami6b01202

Section 3

Nano Biomaterials

Chapter 5

A Facile Method for Formulation of Atenolol Nanocrystal Drug with Enhanced Bioavailability

Luis Castañeda

Abstract

Atenolol is a commonly used antihypertensive drug of class III BCS category. The objective of the present study is to enhance the permeability of atenolol by using a suitable technique which is economical and devoid of using any organic solvents. The nanocrystal technology by high pressure homogenization was chosen for this purpose, which is less expensive and simple method. In this technique, no organic solvent was used. The study was further aimed to characterize prepared nanocrystals in solid state by Fourier-transform infrared spectroscopy (FTIR), powder X-ray diffraction (PXRD) patterns, particle size, zeta potential, % yield, and drug permeation study through isolated goat's intestine. An in vivo study was carried out to determine the pharmacokinetic property in comparison to pure drug powder using rats as experimental animals. The formulation design was optimized by a 3(2) factorial design. In these designs, two factors, namely surfactant amount (X1) and speed of homogenizer (X2), were evaluated on three dependent variables, namely particle size (Y1), zeta potential (Y2), and production yield (Y3).

Keywords: atenolol, nanocrystal, factorial design, ANOVA, antihypertensive, pharmacokinetic

1. Introduction

A nanocrystal is a particle having one or more dimensions of the order of 200 nm or less and considered to have novel characteristics which differentiate them from other materials [1]. When the size of the material is reduced to less than 200 nanometers, the realm of quantum physics takes over and five materials begin to demonstrate entirely new properties. Hence, nanodesign of drugs by various different techniques, like melting, homogenization, and controlled precipitation, is explored to produce drug nanocrystals, nanoparticles, nanosuspensions, etc. [2]. As decrease in size will increase the solubility of drugs, this technology is explored to increase oral bioavailability of sparingly water soluble drugs [3]. Development of soluble and/or permeable drug molecules using nanocrystal formulations has been proven to be successful due to their unique size range and higher surface: volume ratio, which results in enhanced drug dissolution, bioavailability and permeability [4].

Atenolol is a selective $\beta 1$ receptor antagonist, a drug belonging to the group of beta blockers, which is used mainly in different cardiovascular diseases [5]. It often suffers from poor bioavailability after oral dosing due to stumpy permeability through GIT [6]. Approximately 50% of an oral dose is absorbed from the gastro-intestinal tract, the remainder being excreted unchanged in the feces. Researchers have been endeavored to increase its permeability and bioavailability by different techniques including osmotic pump, cyclodextrin-based delivery systems, hydrophilic matrices, transdermal delivery systems, and so on [7–12].

In the present study, we had prepared nanocrystals of atenolol to improve its permeability and modify its solubility, because this method is less time-consuming, required no organic solvents or harsh chemicals like other nanodelivery systems, has a high product yield, has good product stability, and is cheap. High pressure homogenization method was employed to prepare nanocrystals [13]. In this method, high pressure was applied on liquid suspension to force it through a gap or narrow channel inside a pipe. Here, the medium was aqueous containing a hydrophilic surfactant SLS to prevent agglomeration of suspended particles and thus it helped in stabilization. The surfactant used in the study also prevented crystal growth (Ostwald ripening) that could change the dissolution and bioavailability of the drug after storage [14].

2. Materials and methods

2.1 Materials

Atenolol was supplied as a gift sample by Haustus Biotech Pvt. Ltd., Himachal Pradesh, India, and sodium lauryl sulfate, manufactured by Krishna Drug and chemical Pvt. Ltd., Gujrat, was supplied by Mahalakshmi Chemicals Ltd., Greater Noida, India. Triple distilled water was used throughout the experiments. All other chemicals were of reagent grade and used without further purification.

2.2 Preparation of atenolol nanocrystals

Atenolol nanocrystals were prepared by high speed homogenization process using sodium lauryl sulfate as a surfactant [15–17]. The nanocrystals were prepared by adding the different compositions of the surfactant sodium lauryl sulfate and stabilizer (PVP K 30) as mentioned in **Table 1**. Atenolol (1000 mg) was dissolved in

Formulation code	Drug: surfactant	Speed in rpm	Particle size (nm)	Zeta potential (mV)	Production yield
F1	2:1	20,000	312.7 ± 2.0	18 ± 0.2	82 ± 1.0
F2	4:1	20,000	296.7 ± 0.6	20 ± 0.4	88 ± 2.0
F3	4:3	20,000	416.2 ± 0.5	16 ± 0.1	84 ± 1.0
F4	2:1	25,000	210.4 ± 1.0	16.5 ± 0.3	72 ± 0.5
F5	4:1	25,000	125.6 ± 0.5	19 ± 0.2	90 ± 0.8
F6	4:3	25,000	552.6 ± 0.7	17 ± 0.3	88 ± 1.0
F7	2:1	15,000	652.6 ± 2.0	18 ± 0.7	64 ± 1.0
F8	4:1	15,000	620.0 ± 2.5	20 ± 0.4	66 ± 2.0
F9	4:3	15,000	590.0 ± 1.8	19 ± 0.5	64 ± 1.0

Table 1.
Formulation composition of atenolol nanocrystal.

80 ml of distilled water with sodium lauryl sulfate which resulted in a solution of 1 g concentration. The whole procedure was operated at 25 ± 2°C. The solution of drug and surfactant was placed under a high-speed homogenizer (T 25 digital ULTRA-TURRAX IKAR Werke Staufen/Germany) at different speeds (15,000–25,000 rpm) for 70 h. The resulting solution was placed in a tray dryer at 60°C to evaporate the solvent. Nanocrystals were collected and evaluated as required.

2.3 Formulation of capsule dosage of atenolol nanocrystals

Atenolol nanocrystals were mixed with the same ratio of lactose powder, which shows no incompatibility with the drug. The mixture of 1:1 ratio of drug (atenolol nanocrystals) and lactose was prepared. Then, 100 mg of this mixture was filled into the capsules.

3. Characterization of nanocrystals

The nanocrystals of atenolol prepared by the abovementioned method was characterized by the following techniques.

3.1 Particle size and zeta potential analysis

Particle size of the prepared nanocrystals was determined using particle size analyzer (Malvern Instruments Ltd.). The prepared nanocrystals were dispersed in dimethyl sulfoxide and placed in cuvettes and the particle size in terms of average diameter (davg) was determined. Zeta potential was calculated by using Zetasizer ZS 90 (Malvern Instrument Ltd. India) [18].

3.2 Scanning electron microscopy

The morphology of the atenolol nanocrystals was examined by scanning electron microscopy (JSM 6390 India). The sample was mounted on to an aluminum stub and sputter coated for 120 s with platinum particles in an argon atmosphere. The coated samples were then scanned and images were analyzed at 500 or 1000 axis [18].

3.3 Fourier-transform infrared (FTIR) analysis

FTIR analysis of pure atenolol, mixture of atenolol and SLS and obtained nanocrystals and lactose was performed in the range of 4000–500 cm^{-1} as thin KBr pellets using FTIR spectrophotometer (Perkin-Elmer BX II). The observed peaks were reported for functional groups.

3.4 X-ray diffraction study (XRD)

The crystallinities of atenolol and atenolol nanocrystals were evaluated by XRD measurement using an X-ray diffractometer (Bruker AXS, 08 Advance). All samples were measured in the 2θ angle range between 3 and 80° and 0.010 step sizes.

3.5 Percentage yield of production

For any formulation, it is always desirable to have a better production yield so that industrial production becomes feasible not only in terms of cost but also in terms of environmental protection. The production yield of prepared nanocrystals was calculated by the following Eq. (1):

$$Percentage\ yield = (B/A) \times 100 \quad (1)$$

where B is the weight percentage of the final product obtained after drying, and A is the initial total amount of atenolol and sodium lauryl sulfate used for the preparation.

3.6 In vitro release studies of atenolol nanocrystals

The dissolution test was performed in the USP type II apparatus. Nanocrystals (100 mg) were accurately weighed and put into the pretreated dialysis membrane and sealed with clips. The release medium was phosphate buffer (pH 6.8) maintained at 37°C with agitation rate set at 50 rpm. The amount of drug was determined spectrophotometrically at λ_{max} = 275 nm against suitable blank using a preconstructed calibration curve [19].

3.7 In vitro release studies of capsule dosage form of atenolol nanocrystals

The dissolution test was performed on the USP type I apparatus. Capsules containing nanocrystals (100 mg) were accurately placed into the basket of dissolution test apparatus. The release medium was phosphate buffer (pH 6.8) maintained at 37°C with agitation rate set at 50 rpm. The amount of drug released was determined spectrophotometrically at λ_{max} = 275 nm.

3.8 In vitro intestinal permeability studies of pure atenolol and atenolol nanocrystals

The permeability studies of pure atenolol and atenolol nanocrystals were carried out using Franz diffusion cell. To check the intra duodenal permeability, the duodenal part of the small intestine was isolated from sacrificed goat and taken for the in vitro diffusion study. Then this tissue was thoroughly washed with cold Ringer's solution to remove the mucous and lumen contents. The sample solutions were injected into the lumen of the duodenum using a syringe, and the two sides of the intestine were tightly closed. Then the tissue was placed in a chamber of organ bath with continuous aeration and at a constant temperature of 37°C. The receiver compartment was filled with 30 mL of phosphate-buffered saline (pH 5.5). The permeability was tested for 60 minutes. The absorbance was measured using a UV-Vis spectrophotometer at a wavelength of 275 nm, keeping the respective blank. The percent diffusion of the drug was calculated against time and plotted on a graph.

3.9 Stability studies of prepared nanocrystals

The prepared nanocrystals were subjected to stability studies. The nanocrystals were placed in stability chambers for a month at different temperatures, like 4, 25, 37, and 60°C. After 1 month, the tested nanocrystals were subjected to FTIR to find the spectra and compare with the standard spectra of nanocrystals.

3.10 In vivo studies

To determine the in vivo pharmacokinetic parameters for optimized nanocrystal formulation, experimental rats were used. This investigation adhered to the Principles of Laboratory Animal Care. Female albino rats (0.20–0.25 Kg) were

Drug: surfactant (X1)	Speed in rpm (X2)
−1	1
0	1
1	1
−1	0
0	0
1	0
−1	−1
0	−1
1	−1
−1 = 2:1	−1 = 15,000
0 = 4:1	0 = 20,000
+1 = 4:3	+1 = 25,000

Table 2.
Design of experiment for 3^2 factorial analysis.

divided in two groups, each containing six. They were fasted overnight and allowed to administer 0.5 mL aqueous dispersion of pure drug and the most successful formulation of nanocrystal (equivalent to 10 mg/mL atenolol) using oral feeding tube. Blood samples of 0.2 mL were withdrawn through the tail vein of rats after 0.5, 1, 1.5, 2, 2.5, 4, 6, and 24 h of sample administration. The withdrawn samples were centrifuged at 5000 rpm for 20 min. The plasma was separated and stored at −20°C until drug analysis was carried out using HPLC analytical method of analysis. The whole process was carried out according to the reported method by Anwar et al. [20].

3.11 Statistical analysis

Independent T-test was used to analyze data of two batches obtained in various experiments at the 0.05 level of significance by Origin 6.0 software. The difference was considered significant at $p \leq 0.05$.

3.12 Experimental design and statistical analysis

In this study, a 32 full factorial experimental design was introduced to optimize the formulation of nanoparticles. Initial studies were undertaken to decide on the factors and their levels in the experimental design. Based on the results obtained in preliminary experiments, surfactant amount and speed of homogenizer were found to be the major variables in determining the particle size and production yield. So, in this design, two factors, namely surfactant amount (X1) and speed of homogenizer (X2), were evaluated each at three levels and suitably coded (**Table 2**). The effect of these factors were evaluated on three dependent variables, namely particle size (Y1), zeta potential (Y2), and production yield (Y3). A total of 9 formulations were prepared with these variables.

For the studied design, the multiple linear regression analysis (MLRA) method was applied using Statistica 10 (StatSoft Inc., USA) software to fit the full second-order polynomial equation with added interaction terms. Polynomial regression results were demonstrated for the studied responses using Eq. (2):

$$Y = b1 + b2X1 + b3X2 + b4X1X2 + b5X12 + b6X22 \tag{2}$$

where Y is the dependent variable and b1 is the arithmetic mean response of the 9 trials. Coefficient b2 is the estimated coefficient for the factor X1, and coefficient b3 is the estimated coefficient for the factor X2. The main effects (X1 and X2) represent the average result of changing one factor at a time from its low to high value. The interaction terms (X1X2) show how the response changes when two factors interact. The polynomial terms (X12 and X22) are included to investigate nonlinearity. The values of correlation coefficients were set to be statistically significant at 95% confidential interval [21]. To analyze the significance level of all these data, ANOVA was used at 95% confidence interval at 0.05 significance level.

4. Results and discussion

4.1 Characterization of drug, excipients, and their interactions by FTIR

FTIR spectroscopy was used to further characterize possible interactions between the drug and the excipients. The FTIR spectra of atenolol and sodium lauryl sulfate and also of the formulated nanocrystals and lactose were obtained at wavelength ranging from 4000 to 400 cm^{-1}. The spectra obtained from FTIR studies confirmed that there was no major shifting, as well as no loss of functional peaks between the spectra of pure atenolol and atenolol nanocrystals. Comparing the spectra of pure atenolol and atenolol nanocrystals, no difference was shown in the position and trend of the absorption bands. All the distinctive groups in the FTIR spectra of atenolol were found in all the spectra of atenolol nanocrystals, like the amide group ($O{=}C{-}NH_2$) extruding from the benzene ring. Apart from the amide functional group, the presence of the conjugating C=C bond in the benzene ring, the methane (CH), methylene (CH_2) methyl (CH_3), and OH functional group were distinctly observed in the IR spectra (amide:1650 cm^{-1}; CH:2880–2900 cm^{-1}; CH_2: 2916–2936 cm^{-1}; CH_3: 2850 cm^{-1}; conjugating C=C: 1640–1610 cm^{-1}; OH:3200–3550 cm^{-1}), thus providing evidence for the absence of any chemical incompatibility between SLS and atenolol.

4.2 Particle size and zeta potential

The particle size of the atenolol nanocrystal formulations shown in **Table 1** showed a narrow size distribution from 125 to 652 nm, where the intensity of 117.8 nm was 93% and that of 652.5 nm was only 7%. The effect of stirring speed had an enormous effect on particle size. Formulation prepared with 25,000 rpm had smaller size as compared to particle prepared with 15,000 rpm. Concentration of SLS also had an effect on the size distribution. Less concentration of SLS yielded smaller size particles. The formulated nanocrystals were positively charged (16–19 mV), which is desirable for good ocular interaction. Formulation F7 was not further considered due to its larger size (more than 650). This may be due to slow stirring speed.

4.3 Production yield

The date of percentage yield of the prepared nanocrystals (**Table 1**) showed that the atenolol-SLS nanocrystals (batch F5), prepared by drug:SLS ratio 4:1, had comparatively higher yield of production (90%). Stirring speed and concentration of SLS also had an effect on the production yield.

4.4 Scanning electron microscopy

The SEM image, as shown in **Figure 1**, of the atenolol nanocrystals revealed that the particles were crystalline in shape. The average size of the atenolol nanocrystals was found to be less than 200 nm, which was further supported by the results of particle size analysis by Zetasizer.

4.5 X-ray diffraction (PXRD) studies

The powder X-ray diffractogram of pure atenolol powder from 5 to 50° 2θ showed numerous distinctive peaks at 2θ degree that indicated a high crystalline content. The samples were scanned for 2 θ values over a range from 5 to 50°C at a scan rate of 10°/min. The PXRD pattern of pure drug and atenolol nanocrystals were compared with regard to peak positions and relative intensities and presence and/or absence of peaks in certain regions. **Figure 2** represents the XRD photograph of different nanocrystal formulation and pure atenolol powder.

4.6 Permeability study

The permeability study showed increased permeability, when the atenolol was converted into the nanocrystals. The diffusion study showed that the % permeability of nanocrystal formulations was much higher as compared to that of the pure drug. The formulation F5 showed the maximum % release of 90.88%, whereas the pure drug showed only 31.22% release **Figure 3**.

4.7 In vitro dissolution studies

The dissolution rate of pure atenolol was very poor and during a 120-min period, 51.64% of drug was released. The reason for the poor dissolution of pure drug could be poor wettability and poor solubility. In vitro release studies revealed that there was a marked increase in the dissolution rate of atenolol, in the range of 78.30–98.28%, from all nanocrystal formulation compared to pure atenolol. The results revealed that the nanocrystals with a ratio of drug to carrier, 4:1, were having a higher dissolution rate in comparison to all other ratios. This could be attributed to the hydrophilic character of the surfactant and to the amorphous state of the drug. Hence, the present study showed that nanocrystal formulation can

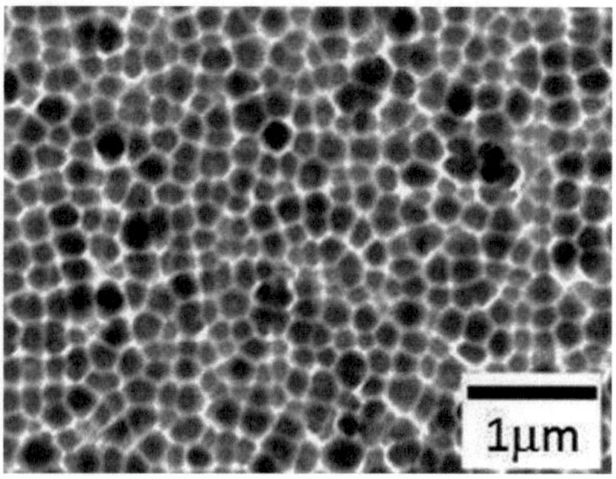

Figure 1.
SEM images of atenolol nanocrystals.

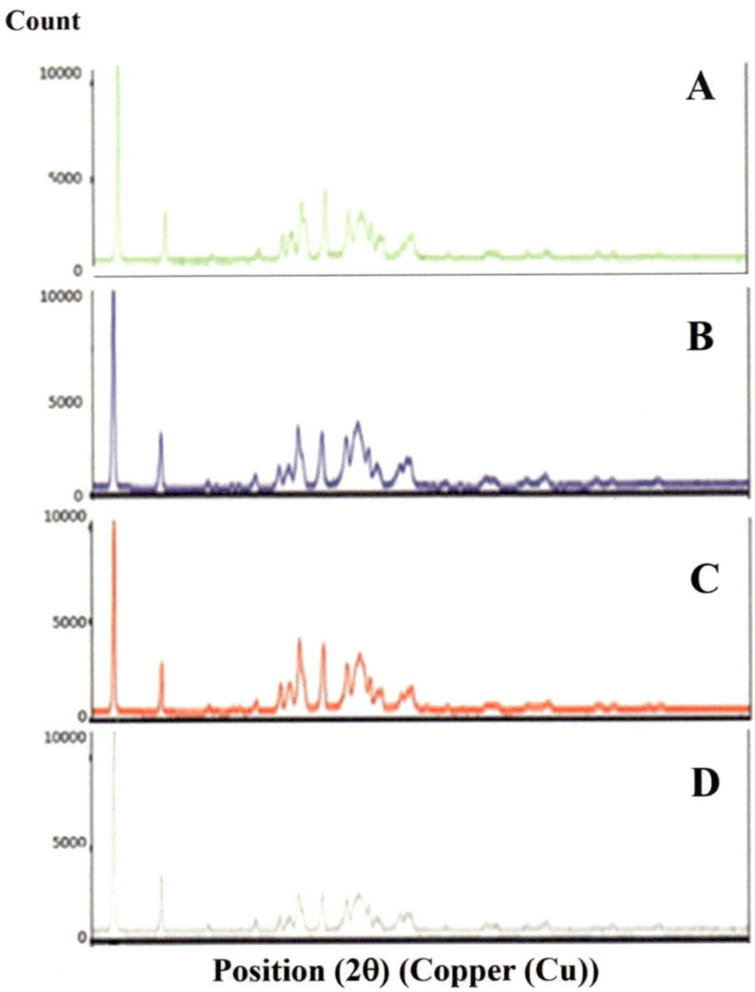

Figure 2.
XRD of atenolol nanocrystals F5 (A), atenolol pure drug (B), atenolol nanocrystals F4 (C), and nanocrystals F1 (D).

be successfully used to enhance dissolution rate of poorly soluble drugs. **Figure 4** shows drug release profile of pure drug nanocrystal formulation [F5] and capsulated nanocrystals.

4.8 Stability studies

The physical appearance of the prepared nanocrystals after keeping them 1 month in stability chambers under various conditions was found to be white to off-white in color, odorless, and crystalline powder. FTIR spectroscopy was used to further characterize possible interactions between the drug and the excipients during the stability studies. There was no major change shown in the FTIR peaks. The prepared nanocrystal formulation was stable during the stability studies done for 1 month.

4.9 In vitro study

The results of in vivo study revealed an improvement in bioavailability of nanocrystal formulation; it was observed that after oral dosing of the drug and the equation 3, their

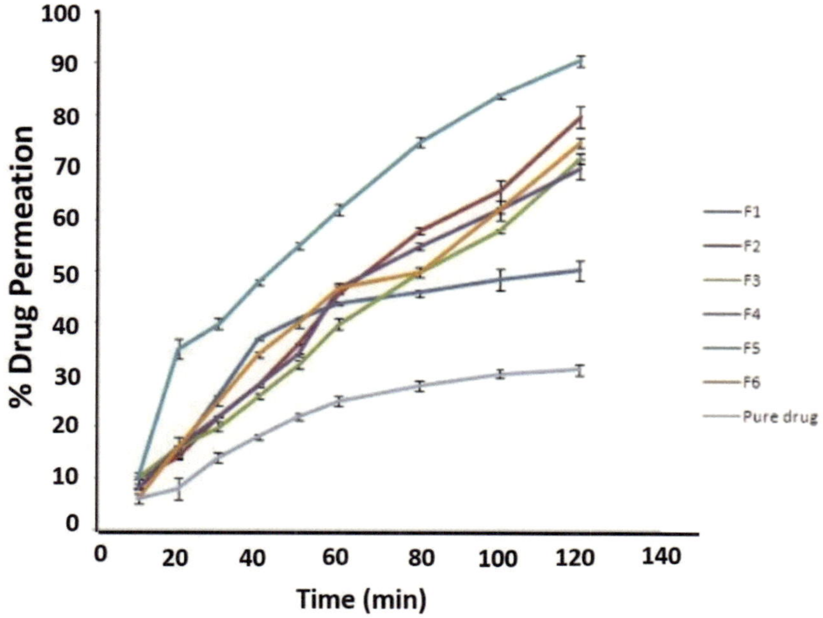

Figure 3.
Permeability from Franz diffusion cell of atenolol pure drug and atenolol nanocrystal formulations.

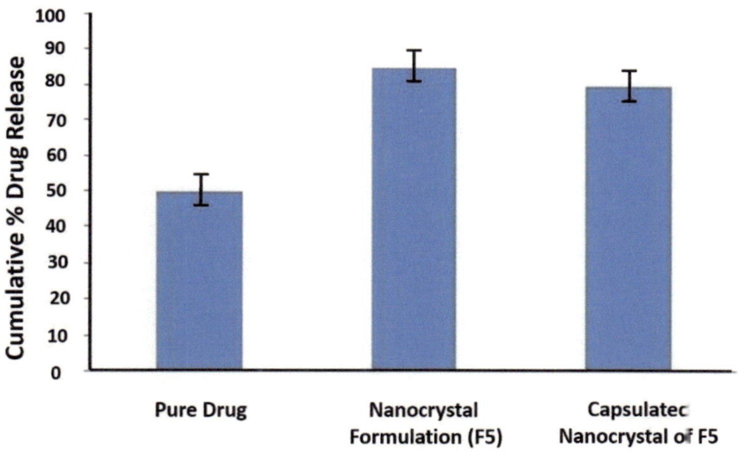

Figure 4.
In vivo dissolution cumulative of % release of atenolol pure drug, atenolol nanocrystals F5, and capsule dosage of nanocrystals in phosphate buffer (pH 6.8).

individual kinetic curve exhibited double peaks. Thus, double peaks can be due to the existence of two absorption sites in the gut interrupted by a region of poor absorption [21]. A rapid attainment of peak plasma concentrations was observed that may be due to the burst release effect brought by the use of SLS for stabilization of nanocrystals. Same phenomenon was reported by Vergote et al. [22]. The AUC0–24 h, MRT, and Cpmax for test formulation were significantly higher at p¡0.05 compared to the drug (**Table 3**).

4.10 Factorial analysis

Applied 3^2 factorial design yields coefficient for one factor and for two factors as well. Coefficient for more than one factor represents interaction of both factors.

Formulation	Cpmax (μg/mL)	Tmax (hour)	AUC0–24 (mAU)	MRT (hour)
Pure drug	612.15 ± 10.6	5.2 ± 1.4	26927.8 ± 4.2	46.56 ± 3.1
Nanocrystal formulation	957.51 ± 20.4	2.8 ± 0.5	75329.3 ± 6.3	84.64 ± 4.6

Table 3.
Pharmacokinetic data of nanocrystal formulation and pure drug.

Coefficient may be positive or negative for synergistic or antagonistic effect, respectively. The coefficients can be directly compared to assess the impact of factors on responses. Obtained polynomial Eqs. (3)–(5) for dependent variables are as follows:

$$\textit{Particles Size} = 4280.7153 - 6169.2601 \times \mathrm{x} - 0.2812 \times \mathrm{y} + 4894.7164 \times \mathrm{x} \times \mathrm{x} + 0.1959 \times \mathrm{x} \times \mathrm{y} + 4.6667E^{-8} \times y \times y \quad (3)$$

$$\textit{Zeta potential} = 31.4304 - 43.3651 \times \mathrm{x} - 0.0004 \times \mathrm{y} + 67.3401 \times \mathrm{x} \times \mathrm{x} - 0.0004 \times \mathrm{x} \times \mathrm{y} + 1E^{-8} \times y \times y \quad (4)$$

$$\textit{Production Yield} = -71.7445 - 370.4401 \times \mathrm{x} + 0.0192 \times \mathrm{y} + 606.0606 \times \mathrm{x} \times \mathrm{x} - 0.0009 \times \mathrm{x} \times \mathrm{y} - 4.2667E^{-7} \times y \times y \quad (5)$$

Speed of homogenizer has a greater effect on the particle size (−0.2812), zeta potential (−0.0004), and production yield (0.0192), whereas amount of surfactant has a lesser effect on the production yield (−370.4401), zeta potential (−43.3651), and particle size (−669.2601).

It was clearly depicted from the magnitude of the coefficients that the amount of surfactant has a positive effect on all the three variables including particle size, zeta potential, and production yield; whereas, the magnitude of the coefficients for speed of homogenizer has an antagonistic effect on all the three variables.

Contour plots and surface plots as shown in **Figures 5** and **6** were plotted, which are very useful to study the interaction effects of the factors on the responses. The response surface depicts the effect of factor contributions at different levels on studied response. Three contour parameters were established for particle size. Drug entrapment and drug release percentage. The contour plots showed very clearly the relationship between the independent variables and the responses.

P value for the effect of drug surfactant ratio is statistically insignificant for particle size and production yield *p* has value greater that 0.05, but it is significant only for zeta potential (p = 0.009851) and particle size (p = 0.035269). Speed of homogenizer has an insignificant effect on zeta potential.

The goodness of fit of the R^2 model was checked by the determination coefficient (R^2). The values of the determination coefficients for particle size (R^2 = 0.78297), zeta potential (R^2 = 0.80392), and production yield (R^2 = 0.8988) indicated that over 95% of the total variations are explained by the model. The values of adjusted determination coefficients (adj R^2 = 0.56594 for particle size, 0.60784 for zeta potential, and 0.79761 for production yield) are also very high (over 90% of the total variations), which indicates a high significance of the model.

A good way to check the model is to enter factor levels from the experimental design (observed response) and generate the predicted response. When we compare the predicted value with actual value, a discrepancy occurs which is called residual.

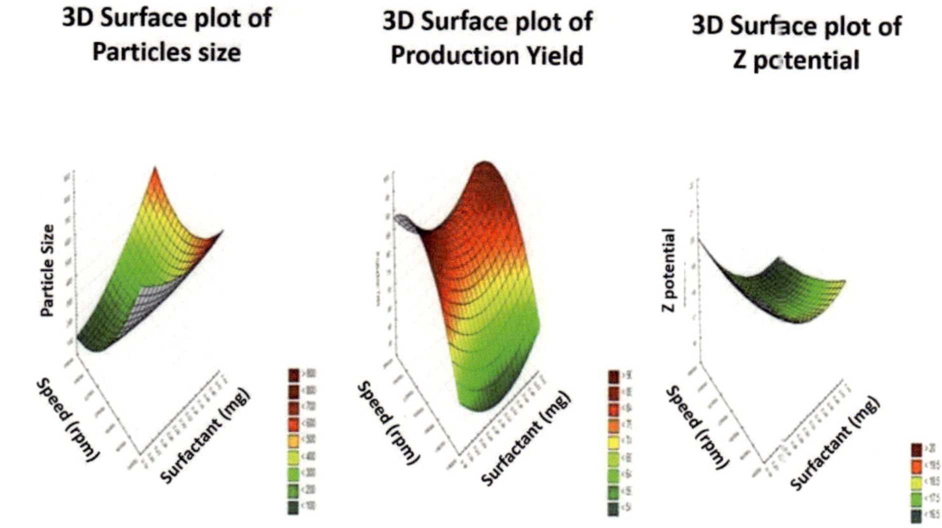

Figure 5.
3D surface plots for the different variables (in order of particle size, production yield, and zeta potential).

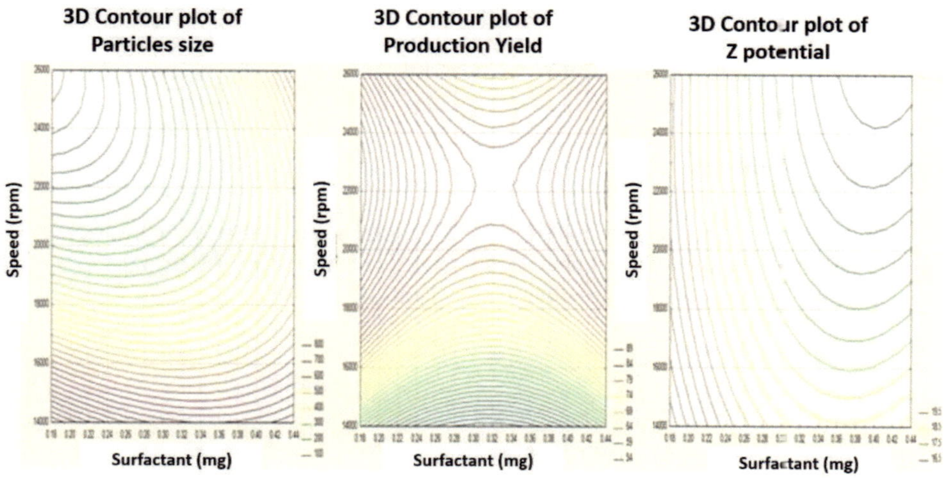

Figure 6.
3D contour plots for the different variables (in order of particle size, production yield, and zeta potential).

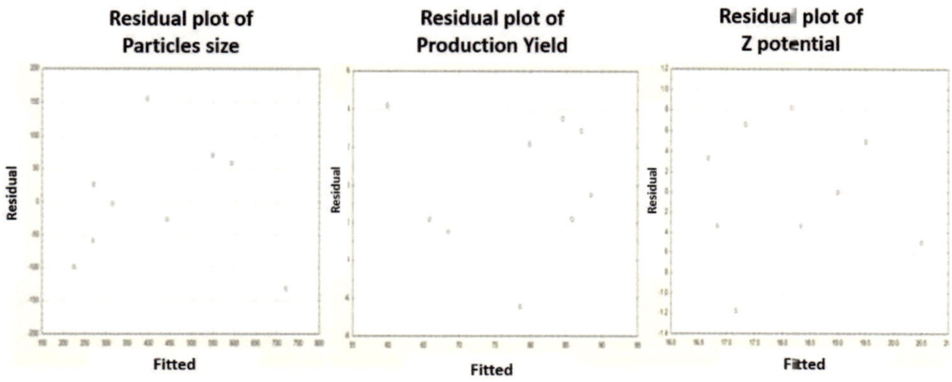

Figure 7.
Plots of residuals for the three different variables (in order of particle size, production yield, and zeta potential).

For statistical purposes, it is assumed that the residuals are normally distributed and independent with constant variance [21]. Residual versus predicted plots were constructed to check the statistical assumptions. In this experiment, there is no definite increase in residuals with predicted levels, which support the underlying statistical assumptions of constant variance. Moreover, the obtained plots of residuals that do not exhibit any systematic structure indicate that the model fits the data well (**Figure** 7). Plots of the residuals versus other predictor variables, or potential predictors that exhibit systematic structure indicate that the form of the function can be improved in some ways.

5. Conclusions

In this study, atenolol nanocrystals had been developed successfully by high-speed homogenization and simultaneous drying, which have shown an improved dissolution and permeability behavior compared to that of the pure drug. Statistical analysis revealed speed of homogenizer had bigger effect on the three observed parameters, whereas amount of surfactant had a lesser effect on them. SEM pictures had shown that the size of the particles obtained after homogenization is below 1 μm. This implied that the size of the drug crystals in these particles was of nanoscale. Therefore, it can be concluded that the selected method of nanocrystal formation and its further optimization by factorial design was effective to increase the solubility, as well as permeability of atenolol.

Acknowledgements

The authors gratefully acknowledge the financial support from the Graduate Studies and Research Section of the Higher School of Medicine, National Polytechnic Institute, through project No. 20190031.

Conflict of interest

The author declares no conflict of interest.

Author details

Luis Castañeda
Sección de Estudios de Posgrado e Investigación de la Escuela Superior de Medicina, Instituto Politécnico Nacional, Plan de San Luis y Díaz Mirón S/N, Casco de Santo Tomas, Cd. de México, Mexico

*Address all correspondence to: lcastaneda@ipn.mx

IntechOpen

© 2019 The Author(s). Licensee IntechOpen. This chapter is distributed under the terms of the Creative Commons Attribution License (http://creativecommons.org/licenses/by/3.0), which permits unrestricted use, distribution, and reproduction in any medium, provided the original work is properly cited. CC BY

References

[1] Fichera MA, Keck CM, Muller RH. Nanopure technology-drug nanocrystals for the delivery of poorly soluble drugs. In particles; 2004

[2] Pace GW. Process to generate submicron particles of water insoluble compounds. US Patent 6-177-103; 2001

[3] Bushrab NF, Muller RH. Nanocrystals of poorly soluble drags for oral administrations. New Drugs. 2003;**5**:20-22

[4] Srivalli KMR, Mishra B. Drugs nanocrystals: A way toward scale-up. Saudi Pharmaceutical Journal. 2016;**24**(4):384-404

[5] Wander GS, Chhabra ST, Kaur K. Atenolol drug profile. Supplement: Journal of Association of Physicians of India. 2009;**57**:13-16

[6] Wu F, Cheni P, Lee YJ, Long RR. Comparative pharmacokinetics of two atenolol products. Journal of Food and Drug Analysis. 2003;**11**(1):4-7

[7] Singh B, Kaur S, Chakkal AN. Formulation and optimization of controlled release mucoadhesive tablets of atenolol using response surface methodology. AAPS PharmSciTech. 2006;**7**:E19-E28

[8] Vaithiyalingam SR, Sastry SV, Dehon RH, Reddy LK, Khan MA. Long-term stability characterization of a controlled release gastrointestinal therapeutic system coated with cellulose acetate pseudolatex. Pharmazie. 2001;**56**:66-69

[9] Kim J, Shin SC. Controlled release of atenolol from the ethylene-vinyl acetate matrix. International Journal of Pharmaceutics. 2004;**273**:23-27

[10] Jug M, Becirevic-Lacan M, Bengez S. Novel cyclodextrin-based film formulation intended for buccal delivery of atenolol. Drug Development and Industrial Pharmacy. 2009;**35**:796-807

[11] Kulkarni A, Bhatia M. Development and evaluation of regioselective bilayer floating tablets of atenolol and lovastatin for biphasic release profile. Iranian Journal of Pharmaceutical Research. 2009;**8**:15-25

[12] Srivastava AK, Wadhwa S, Ridhurkar D, Mishra B. Oral sustained delivery of atenolol from floating matrix tablets-formulation and in vitro evaluation. Drug Development and Industrial Pharmacy. 2005;**31**:367-374

[13] Kobierski S, Hanisch J, Mauludin R, Muller RH, Keck C. Nanocrystal production by smartcrystal combination technology. International Symposium on Controlled Release of Bioactive Materials. 2008;**35**:3239

[14] Loh ZH, Samanta AK, Heng PWS. Overview of milling techniques for improving the solubility of poorly water-soluble drugs. Asian Journal of Pharmaceutical Sciences. 2015;**19**:255-274

[15] Merisko-Liversidge E, Liversidge GG, Copper ER. Nanosizing: A formulation approach for poorly-water-soluble compounds. European Journal of Pharmaceutical Sciences. 2003;**18**:113-120

[16] Akkar A, Muller RH. Nanocrystals of itraconazole and amphotericin B produced by high pressure homogenization. In: Annual Meeting of the American Association of Pharmaceutical Scientists. Valencia/USA: American Scientific Publishers; 2003. pp. 479-491

[17] Mauludin R, Möschwitzer J, Müller RH. Comparison of ibuprofen drug nanocrystals produced by high

pressure homogenization (HPH) versus ball milling. AAPS Annual Meeting. Nashville; 2005. p. T2217

[18] Singh A, Deep A. Formulation and evaluation of nanoparticles containing atenolol. International Journal of Pharmaceutical Research. 2011;**3**:59-62

[19] Moneghini M, Carcano A, Zingone G, Perissutti B. Studies in dissolution enhancement of atenolol. International Journal of Pharmaceutics. 1998;**175**:177-183

[20] Anwar M, Warsi MH, Mallick N, Akhter S, Gahoi S, Jain GK, et al. Enhanced bioavailability of nano-sized chitosan-atorvastatin conjugate after oral administration to rats. European Journal of Pharmaceutical Sciences. 2011;**44**:241-249

[21] Mummaneni V, Dressman JB, Amidon GL. Gastric pH influences the appearance of double peaks in the plasma concentration-time profiles of cimetidine aster administration in dogs. Pharmaceutical Research. 1995;**12**:780-786

[22] Vergote GJ, Vervaet C, Van Driessche I, Hoste S, De Smedt S, Demeester J, et al. In-vivo evaluation of matrix pellets containing nanocrystalline ketoprofen. International Journal of Pharmaceutics. 2002;**240**:79-84

Chapter 6

Production, Processes and Modification of Nanocrystalline Cellulose from Agro-Waste: A Review

R.A. Ilyas, S.M. Sapuan, R. Ibrahim, M.S.N. Atikah, A. Atiqah, M.N.M. Ansari and M.N.F. Norrrahim

Abstract

Nanocrystalline cellulose is a renewable nanomaterial that has gained huge attention for its use in various applications from advanced biomedical material to food packaging material due to its exceptional physical and biological properties, such as high crystallinity degree, large specific surface area, high aspect ratio, high thermal resistance, good mechanical properties, abundance of surface hydroxyl groups, low toxicity, biodegradability, and biocompatibility. However, they still have drawbacks: (1) sources of raw materials and its utilization in the production of nanocomposites and (2) high chemical and energy consumption regarding the isolation of macro-sized fibers to nano-sized fibers. The incorporation of hydrophilic nanocrystalline cellulose within hydrophobic polymer limits the dispersion of nano-sized fibers, thus resulting in low mechanical properties of nanocomposites. Hence, surface modification on nano-sized fiber could be a solution to this problem. This review focuses on the advanced developments in pretreatment, nanocrystalline production and modifications, and its application in food packaging, biomedical materials, pharmaceutical, substitution biomaterials, drug excipient, drug delivery automotive, and nanopaper applications.

Keywords: nanocrystalline cellulose, nanocomposites, surface modification, hydrolysis, agro-waste

1. Introduction

During the past decades, huge efforts have been made to improve new chemicals and/or materials and replace broadly used petroleum-based products by utilizing biomass renewable feedstock [1–3]. Biocompatible composites and biodegradable plastics produced from biorenewable resources are regarded as promising biomaterials that could replace petrochemical-based polymers and hence reduce global dependence on nonrenewable sources (i.e., fossil fuels: coal, petroleum, and natural gas) and provide simplified recycling or end-of-life disposal [4–10].

Agro-based industry's function is to increase the value of raw agricultural products through downstream processing so that products are marketable, consumable, and used to generate income and provide profit to the producer [11]. However, there is waste generated through the process of downstream and upstream of agro-industry. The composition of industrial wastes varies depending on the types of industry as different countries apply various categories for industrial waste which contribute adversely to air, soil, and water quality. This is due to some of the industrial wastes which are neither toxic nor hazardous. For example, organic wastes, such as corncob, sugarcane bagasse, sugar palm (fiber, frond, bunch, trunk), areca nut husk fiber, wheat straw fiber, soy hull fiber, pineapple leaf fiber, oil palm (mesocarp fiber, empty fruit bunch, frond), rubber wood thinning, curaua fiber, banana fiber, water hyacinth fiber, wheat straw, sugar beet fiber, etc. that are produced by agro-based industries are not hazardous in nature and thus have potential for other uses [12–14]. **Figure 1** shows the by-products of agro-industry that are used for sources of lignocellulose biomass.

Biomass renewable feedstocks are of great interest due to the possibility of nontoxicity, renewability, and biodegradability as well as sustainability [12–17]. Lignocellulosic can be classified as lower-value biomass (LVB). Lower-value biomass (LVB) in forest or agriculture industry constitutes noncommercial material traditionally left on site following harvesting of crops. However, emerging markets for energy, chemicals, and bioproducts have increased incentives to harvest and utilize this material in some cases [20–25]. Lignocellulosic biomass suppliers do not use any kind of wood indiscriminately due to economic and environmental reasons; they usually used mobilized woody biomass sourced from by-products of forest operations, agriculture, and crops' waste as well as the wood industry waste such as sawmills. Lignocellulosic biomass sector has been developed to work in synergy with other agro-based industry and wood-based industries to give value to non-mobilized and/or low-value biomass such as trunk, fiber, sugar cane bagasse, manure bedding, plant stalks, vines, hulls, leaves, vegetable matter, sawdust, mill

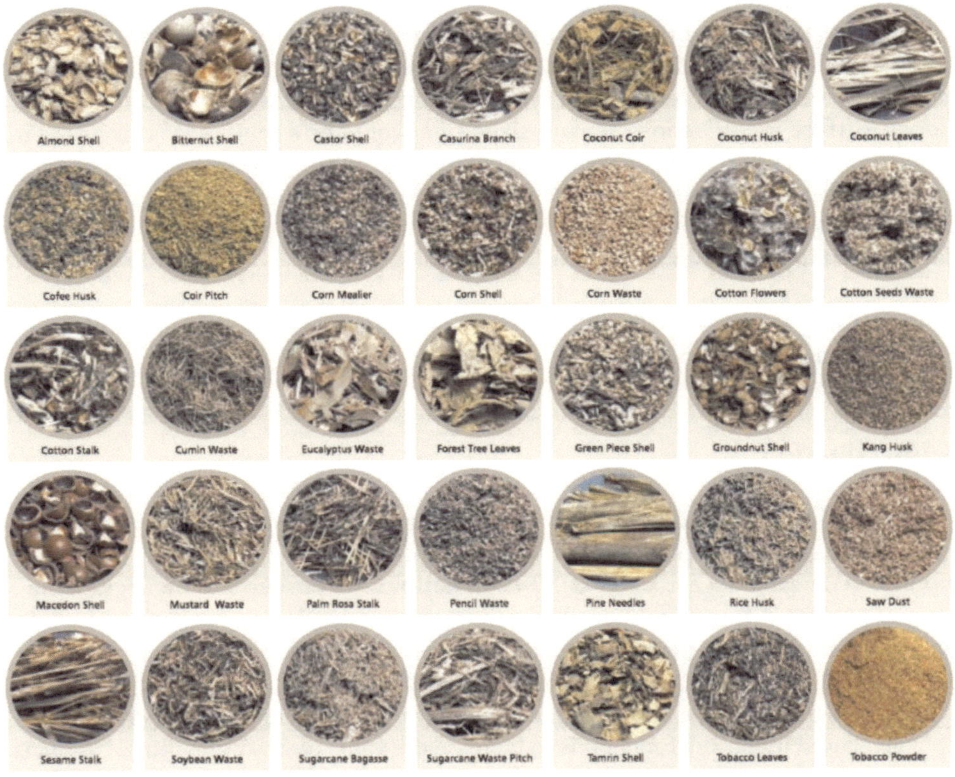

Figure 1.
By-products of agro-industry that are used for sources of lignocellulose biomass.

Figure 2.
By-products of forest operation that are used for sources of lignocellulose biomass. Adapted from Ref. [23]. http://www.europeanbioenergyday.eu/solid-bioenergy-in-questions-an-asset-to-eu-forests/.

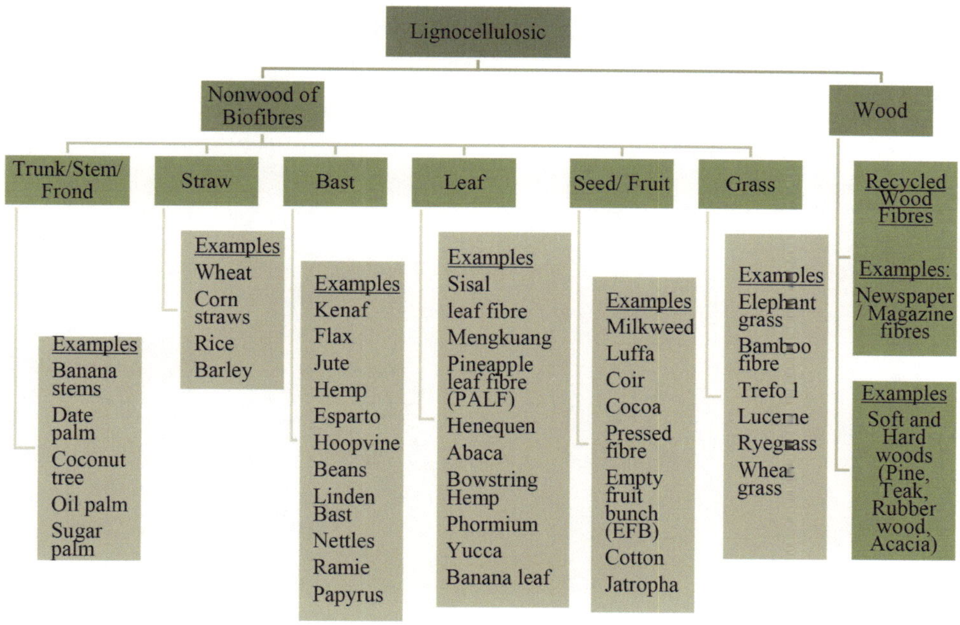

Figure 3.
Schematic representation of lignocellulosic agro-waste and by-product of forest classification. Adapted from Ref. [7].

residues, thinnings, low-quality wood, tops, and limbs. Biomass generators do not use high-quality timber or main agricultural products, as using lumber or major crops would make the price of biomass wholly uncompetitive for end consumers. **Figure 2** shows the by-products of forest operation that are used for sources of lignocellulose biomass. Natural fibers or lignocellulosic fibers can be classified into two main groups that are wood and non-wood bio-fibers (**Figure 3**). This review will be focusing on production, processes, modification, and application of nanocrystalline cellulose from agro-waste.

2. Lignocellulosic biomass from agro-waste fiber and forest by-products

Lignocellulosic biomass comprises of three major chemical components that are cellulose, lignin, and hemicellulose [18–21]. The chemical compositions of

Fibers	Holocellulose (wt%)		Lignin (wt%)	Ash (wt%)	Extractives (wt%)	Crystallinity (%)	Ref.
	Cellulose (wt%)	Hemicellulose (wt%)					
Sugar palm fiber	43.88	7.24	33.24	1.01	2.73	55.8	[6]
Sugar palm frond	66.49	14.73	18.89	3.05	2.46	—	[28]
Sugar palm bunch	61.76	10.02	23.48	3.38	2.24	—	[28]
Sugar palm trunk	40.56	21.46	46.44	2.38	6.30	—	[28]
Wheat straw fiber	43.2 ± 0.15	34.1 ± 1.2	22.0 ± 3.1	—	—	57.5	[29]
Soy hull fiber	56.4 ± 0.92	12.5 ± 0.72	18.0 ± 2.5	—	—	59.8	[29]
Areca nut husk fiber	34.18	20.83	31.60	2.34	—	37	[14]
Helicteres isora plant	71 ± 2.6	3.1 ± 0.5	21 ± 0.9	—	—	38	[30]
Pineapple leaf fiber	81.27 ± 2.45	12.31 ± 1.35	3.46 ± 0.58	—	—	35.97	[31]
Ramie fiber	69.83	9.63	3.98	—	—	55.48	[32]
Oil palm mesocarp fiber (OPMF)	28.2 ± 0.8	32.7 ± 4.8	32.4 ± 4.0	—	6.5 ± 0.1	34.3	[33]
Oil palm empty fruit bunch (OPEFB)	37.1 ± 4.4	39.9 ± 0.75	18.6 ± 1.3	—	3.1 ± 3.4	45.0	[33]
Oil palm frond (OPF)	45.0 ± 0.6	32.0 ± 1.4	16.9 ± 0.4	—	2.3 ± 1.0	54.5	[33]
Oil palm empty fruit bunch (OPEFB) fiber	40 ± 2	23 ± 2	21 ± 1	—	2.0 ± 0.2	40	[34]
Rubber wood	45 ± 3	20 ± 2	29 ± 2	—	2.5 ± 0.5	46	[34]
Curaua fiber	70.2 ± 0.7	18.3 ± 0.8	9.3 ± 0.9	—	—	64	[35]
Banana fiber	7.5	74.9	7.9	0.01	9.6	15.0	[36]
Sugarcane bagasse	43.6	27.7	27.7	—	—	76	[37]
Kenaf bast	63.5 ± 0.5	17.6 ± 1.4	12.7 ± 1.5	2.2 ± 0.8	4.0 ± 1.0	48.2	[38]
Phoenix dactylifera palm leaflet	33.5	26.0	27.0	6.5	—	50	[39]

DOI: http://dx.doi.org/10.5772/intechopen.87001

Fibers	Holocellulose (wt%)		Lignin (wt%)	Ash (wt%)	Extractives (wt%)	Crystallinity (%)	Ref.
	Cellulose (wt%)	Hemicellulose (wt%)					
Phoenix dactylifera palm rachis	44.0	28.0	14.0	2.5	—	55	[39]
Kenaf core powder	80.26	23.58	—	—	—	48.1	[40]
Water hyacinth fiber	42.8	20.6	4.1	—	—	59.56	[41]
Wheat straw	43.2 ± 0.15	34.1 ± 1.2	22.0 ± 3.1	—	—	57.5	[42]
Sugar beet fiber	44.95 ± 0.09	25.40 ± 2.06	11.23 ± 1.66	17.67 ± 1.54	—	35.67	[43]
Mengkuang leaves	37.3 ± 0.6	34.4 ± 0.2	24 ± 0.8		2.5 ± 0.02	55.1	[44]

Table 1.
Chemical composition of agro-waste fibers and forest by-products from different plants and different parts.

	Cellulose	Hemicellulose	Lignin
Structure	• Cellulose is assembled together with pectin fibers, which function to bind the cellulose together to produce tighter cell walls in natural fibers, accounting for their strength providing resistance to lysing in the presence of water • Hemicelluloses consist of long chain—7000–15,000 glucose molecules per polymer	• Hemicellulose is a cell wall polysaccharide that has the capacity to bind strongly to cellulose microfibrils by hydrogen bonds • Hemicelluloses consist of short chains—500–3000 glucose molecules per polymer	• Lignin is a cross-linked polymer with molecular masses in excess of 10,000 u
Function	• Connecting cells to form tissue • Provide structural support • Provides a strong resistance to stress • Prevents the cell from bursting in hypotonic solution	• Responsible for the moisture absorption, biodegradation • Microfibrils are cross-linked together by hemicellulose homopolymers	• Responsible for UV degradation • Lignin assists and strengthens the attachment of hemicelluloses to microfibrils • Lignin plays a crucial part in conducting water in plant stems
Properties	• Thermal stability (occurred from 315 to ~400°C)	• Thermal stability occurred from 220 to ~315°C	• Thermal stability occurred from 165 to ~900°C

Table 2.
Functions and properties of cellulose, hemicellulose, and lignin. Adapted from Refs. [6, 7, 27].

agro-waste fibers are different depending on the type of fiber as summarized in **Table 1**. Besides that, it can be concluded in **Table 1** that the highest cellulose contents are pineapple leaf fibers (81.27%), followed by kenaf core powder (80.26%). Besides that, from **Table 1** also we can summarize that the chemical composition of natural fibers is 30–80% cellulose, 7–40% hemicellulose, and 3–33% lignin. Cellulose, hemicellulose, and lignin have their own properties and functionality. **Table 2** shows the functional properties of the cellulose, hemicellulose, and lignin. The physical, thermal, and mechanical properties of the natural fibers are diverse between each other as they are mostly depending on cellulose crystallinity. Intra- and intermolecular hydrogen bonding among the cellulose chains affects the packing compactness of cellulose crystallinity. **Table 1** shows the chemical composition of natural fibers and their crystallinity. From the abovementioned lignocellulosic, particularly, the hemicellulose and cellulose have promising features such as existing refining agro-forest or agro-waste factories. For centuries, cellulose has been utilized in the form of non-wood plant fibers and wood as building materials, clothing, textile, and paper.

3. Nanocrystalline cellulose

Nanocrystalline cellulose (NCC) has several notable optical, chemical, and electrical properties due to their needlelike shape, high surface area, high aspect ratio (length/diameter), high crystallinity, nanoscale size, high strength and stiffness, low density, and highly negative charge which lead to unique behavior in solutions. The high chemical reactivity of the surface makes NCC customizable for various applications, besides their heat stability which allows high-temperature applications. Moreover, they also have huge surface OH groups which provide active sites for hydrogen bonding through the interlocking with nonpolar matrix [4, 7, 10, 45, 46]. Nanocrystalline cellulose can be isolated from cellulose as shown in **Figure 4**. The nanocellulose can be obtained through two approaches: top-down by the disintegration of plant fiber or bottom-up by biosynthesis [46]. For bottom-up biosynthesis approach, fermentation of low-molecular-weight sugars occurred by using bacteria from *Acetobacter* species. Meanwhile, for the top-down approach, the production of nanocrystalline cellulose is chemically induced via removing amorphous region. The chemical or mechanical treatments or a combination of both treatments involves enzymatic treatment, grinding, high-pressurized homogenization, acid hydrolysis, TEMPO-mediated oxidation, microfluidization, cryocrushing, and high-intensity ultrasonification. **Table 3** shows the hydrolysis approaches from various sources of agro-waste and forest by-product for NCC isolation.

Figure 4.
Schematic representation of lignocellulosic agro-waste and by-product of forest classification. Adapted from Ref. [47].

Source	Process	References
Acacia mangium	H_2SO_4 hydrolysis	[56]
Algae	H_2SO_4 hydrolysis	[57]
Areca nut husk fiber	HCl hydrolysis	[14]
Bacterial cellulose	H_2SO_4 hydrolysis	[58]
Bamboo	H_2SO_4 hydrolysis	[59]
Bamboo (*Pseudosasa amabilis*)	H_2SO_4 hydrolysis	[60]
Banana fiber	$H_2C_2O_4$ hydrolysis	[31]
Banana pseudo-stem	TEMPO-mediated oxidation, formic acid hydrolysis	[61]
Cassava bagasse	H_2SO_4 hydrolysis	[62]

Source	Process	References
Coconut husk	H_2SO_4 hydrolysis	[63]
Colored cotton	H_2SO_4 hydrolysis	[64]
Corncob	H_2SO_4 hydrolysis	[13]
Cotton (cotton wool)	H_2SO_4 hydrolysis	[65]
Cotton linters	HCl hydrolysis	[66]
Cotton Whatman filter paper	H_2SO_4 hydrolysis	[67]
Cotton (*Gossypium hirsutum*) linters	H_2SO_4 hydrolysis	[68]
Cotton stalk	TEMPO-mediated oxidation and H_2SO_4 hydrolysis	[69]
Cotton fiber	H_2SO_4 hydrolysis	[70]
Curaua fiber	H_2SO_4, H_2SO_4/HCl, HCl hydrolysis	[35]
Eucalyptus kraft pulp	H_2SO_4 hydrolysis	[71]
Grass fibers	H_2SO_4 hydrolysis	[72]
Grass fibers (*Imperata brasiliensis*)	H_2SO_4 hydrolysis	[73]
Groundnut shells	H_2SO_4 hydrolysis	[74]
Hibiscus sabdariffa fibers	Steam explosion H_2SO_4 hydrolysis	[75]
Humulus japonicus stem	H_2SO_4 hydrolysis with high-temperature pretreatment	[76]
Industrial bioresidue	H_2SO_4 hydrolysis	[77]
Industrial bioresidue (sludge)	H_2SO_4 hydrolysis	[78]
Kraft pulp	H_2SO_4 hydrolysis	[79]
Kenaf core wood	H_2SO_4 hydrolysis	[40]
MCC	H_2SO_4 hydrolysis	[55]
Mengkuang leaves	H_2SO_4 hydrolysis	[44]
Mulberry	H_2SO_4 hydrolysis	[80]
Oil palm trunk	H_2SO_4 hydrolysis	[81]
Oil palm empty fruit bunch (OPEFB)	H_2SO_4 hydrolysis	[82]
Phormium tenax (harakeke) fiber	H_2SO_4 hydrolysis	[83]
Potato peel waste	H_2SO_4 hydrolysis	[84]
Flax fiber	H_2SO_4 hydrolysis	[83]
Ramie	KOH hydrolysis	[85]
Ramie	H_2SO_4 hydrolysis	[86]
Ramie	H_2SO_4 hydrolysis	[87]
Rice husk	H_2SO_4 hydrolysis	[63]
Rice straw	H_2SO_4 hydrolysis	[88]
Sesame husk	H_2SO_4 hydrolysis	[89]
Sisal fiber	H_2SO_4 hydrolysis	[90]
Soy hulls	H_2SO_4 hydrolysis	[91]
Sugar palm fiber	H_2SO_4 hydrolysis	[6]
Sugar palm frond	H_2SO_4 hydrolysis	[92]

Source	Process	References
Sugarcane bagasse	H_2SO_4 hydrolysis	[37]
Sago seed shells	H_2SO_4 hydrolysis	[93]
Tunicate	H_2SO_4 hydrolysis	[94]
Water hyacinth fiber	HCl hydrolysis	[48]
Wood pulp	TEMPO oxidation followed by HCl hydrolysis	[95]
Wheat straw	H_2SO_4 hydrolysis	[96]
Valonia ventricosa	HCl hydrolysis	[97]

Table 3.
Available process of extraction approaches from different sources for NCC isolation.

4. Processes of nanocrystalline cellulose

Recently, researchers are exploring the potential utilization of agriculture or forest wastes as NCCs' sources. As a consequence, the various local sources are used to investigate the potential of NCC in certain technologies. The isolation of NCC needs intensive hydrolysis chemical treatment. However, according to the degree of processing and raw material, physical, chemical, enzymatic, and ionic pretreatments are performed before nanocrystalline cellulose synthesis. **Figure 5** shows the sources, pretreatments, synthesis, and application of nanocrystalline cellulose. It is good to know that appropriate pretreatments of cellulosic fibers promote the accessibility of hydroxyl group, alter crystallinity, increase the inner surface, and break cellulose hydrogen bonds and hence improved the reactivity of the fibers [6, 7, 10]. Several approaches to diminish cellulosic fibers into nanofibers can be divided into several techniques such as acid hydrolysis, alkali treatment, mechanical treatments, and combination of mechanical and chemical treatments. Common methods for isolate NCC are hydrolysis methods which are a chemical method. **Figure 6** shows

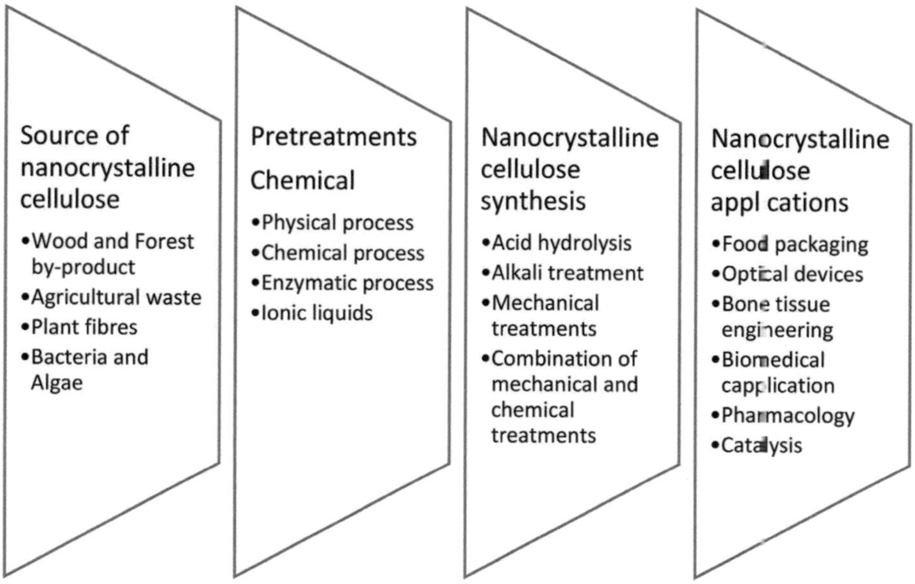

Figure 5.
Sources, pretreatments, synthesis, and application of nanocrystalline cellulose. Adapted from Refs. [6, 7, 10].

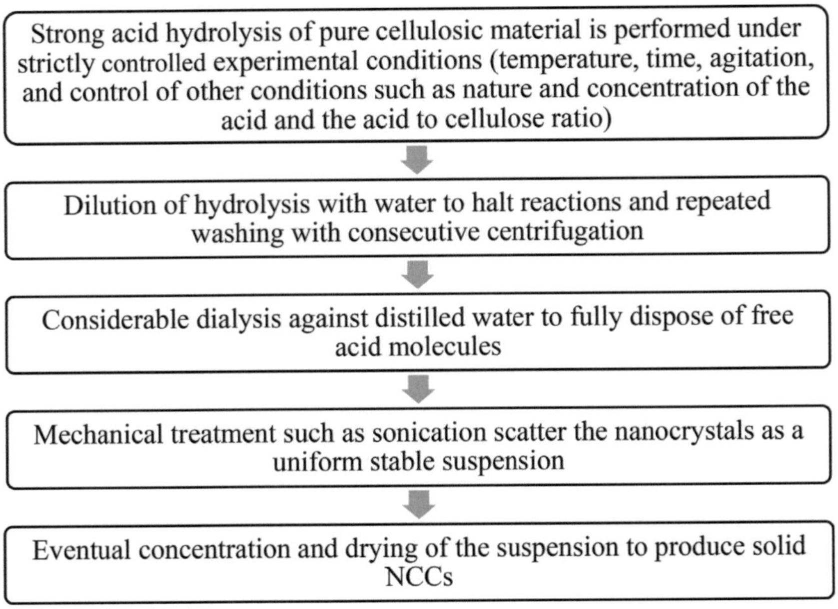

Figure 6.
Typical process for the production of nanocrystalline cellulose. Adapted from Refs. [4, 5, 8].

the typical process for the production of nanocrystalline cellulose. Hydrolysis process includes inserting raw plant fibers into a strong acidic environment with the help of mechanical agitation. Concentrated acid and shear forces on solution generate shear rates in the stream and decrease the size of fibers to the nanoscale. Sulfuric acid (H_2SO_4) is commonly used in the isolation process of NCC besides other chemicals such as HCL [48], HBr [49], and H_3PO_4 [50]. Hydrolysis process using sulfuric acid solution resulted in a high number of negatively charged sulfate groups on the surface of NCC. This process limits the agglomeration and flocculation of NCC in an aqueous medium [51]. The drawback from this process is that the NCC displays moderate thermostability. Hence to overcome this drawback, the NCC will either undergo dialysis process using distilled water to fully dispose free acid molecules or use sodium hydroxide (NaOH), which functions to neutralize nanoparticles [52]. **Figure** 7 displays three steps in the mechanism of acid hydrolysis [53]:

1. Development of conjugated acid by reactions between oxygen protons and glycoside acid

2. Breaking down of C-O bonds and segregation of conjugated acid into cyclic carbonium ions

3. Release of the proton and free sugar after the addition of water

There are numerous studies that have been conducted on the effects of concentration of acid, acid-to-fiber ratio, and temperature and time of the hydrolysis process on the dimensions and morphological properties of yielded nanocrystalline cellulose. According to Azizi et al. [29], there is a strong relationship between the hydrolysis time and acid-to-fiber ratio to the length and dimensions of nanocrystalline cellulose, which by increasing the hydrolysis time and acid-to-fiber ratio would reduce the dimension and length of nanocrystalline cellulose.

Besides that, there are large numbers of published studies [51, 54] that describe the dimension, size, and shape of NCC that were affected by the conditions of hydrolysis process (purity of the material, temperature, time, and ultrasound

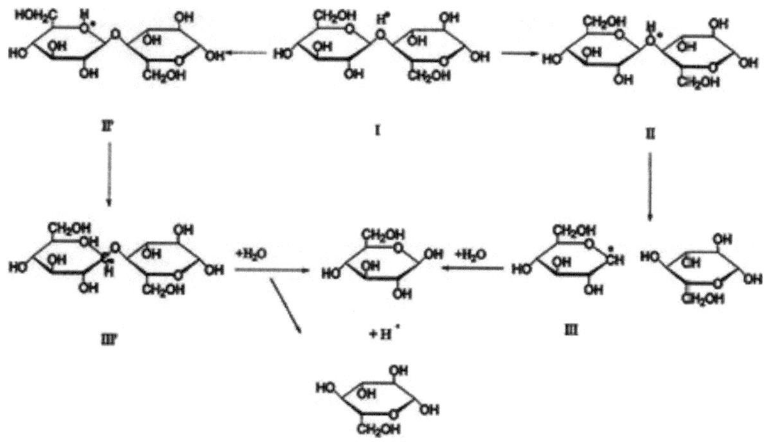

Figure 7.
Mechanism of hydrolysis of acid [53].

treatment) and a variety of cellulosic fiber sources. Bondeson et al. [55] conducted an experiment on the isolation of NCC and found that the optimized condition is at a concentration of 63.5% H_2SO_4, which yielded 38 wt.% of NCCs with a width of 10 nm. Another experiment that is conducted by Ilyas et al. [6] found that the optimum yield for isolating sugar palm nanocrystalline cellulose is at a concentration of 60 wt% H_2SO_4 and duration hydrolysis of 45 min, with length and diameters of 130 ± 30 and 9 ± 1.96 nm, respectively. **Table 3** shows the preparation of NCC using various acid hydrolysis processes from different cellulosic sources. Typical procedures for NCC extraction are composed of several steps: strong acid hydrolysis, dilution, dialysis, sonification, and drying of NCC.

5. Limitation and modification of nanocrystalline cellulose

There are several limitations when using natural fibers as reinforcement filler in the polymer matrix such as single-particle dispersion, barrier properties,

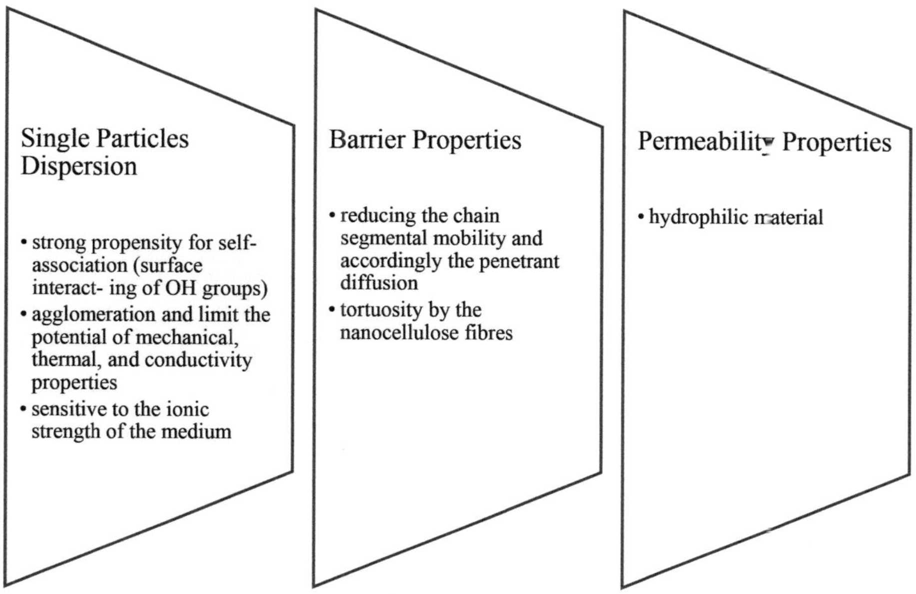

Figure 8.
Limitation of nanocellulose. Adapted from Refs. [6, 7, 10].

permeability properties, and poor interfacial adhesion (**Figure 8**). Nanocrystalline cellulose has a strong propensity of self-association due to the interaction of abundance OH groups within its surface, which causes agglomeration and limits its potential applications. Besides, hydrophilic properties of nanocrystalline cellulose make it difficult to disperse homogenously within any medium and matrix. Therefore, in order to overcome the incompatible nature, poor interfacial adhesion, and difficult dispersion of nanocrystalline cellulose in a polymer matrix, surface modification of fibers or modification of matrix is introduced. Nanocellulose displays a high surface area valued more than 100 m^2/g. This gives advantages to nanocellulose for surface modification in order to introduce any desired surface functionality. However, according to Postek et al. [98], the surface chemistry of nanocellulose is primarily controlled by the process of isolation that used to prepare these nanocelluloses from raw cellulose substrate. **Figure 9** shows the most common surface chemical modifications of nanocrystalline cellulose. Surface modification of NCC can be categorized into three typical groups, namely, (1) polymer grafting based on "grafting onto" strategy with different coupling agents (as indicated with blue arrows in **Figure 9**), (2) substitution of hydroxyl group with small molecules (as indicated with red arrows in **Figure 9**), and (3) polymer grafting

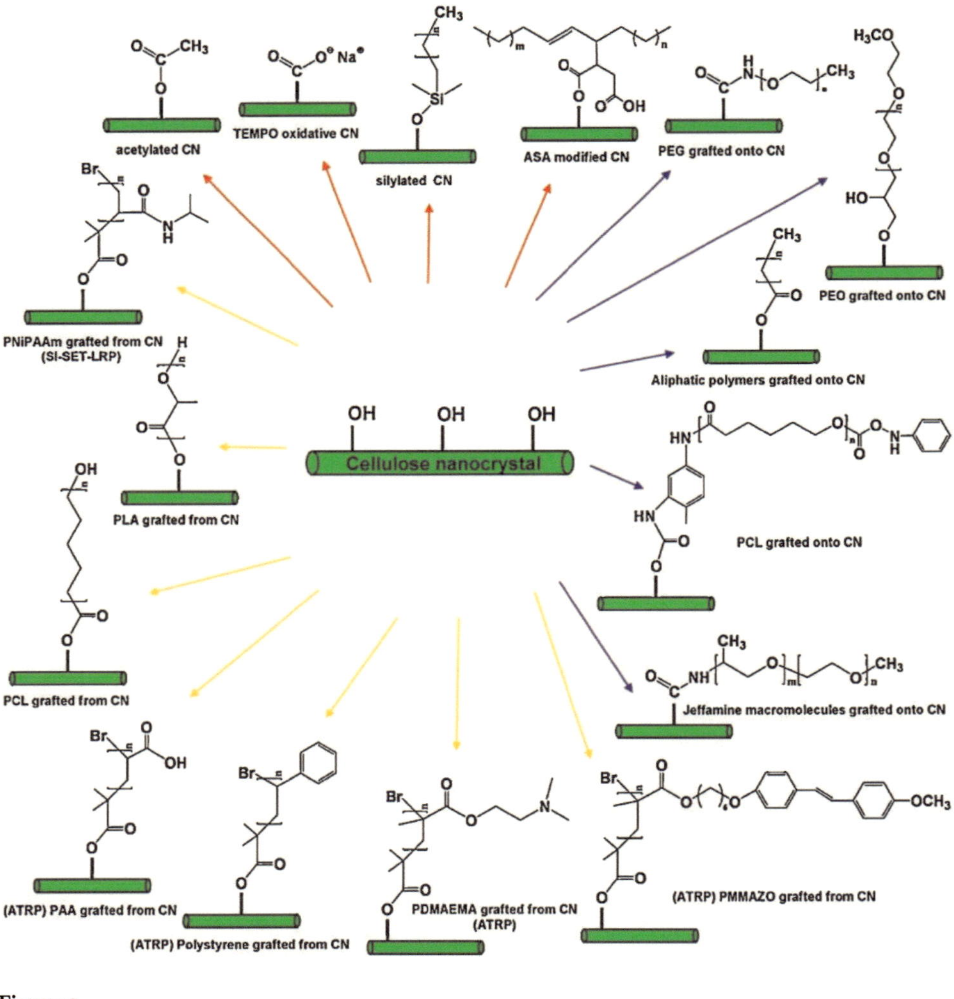

Figure 9.
Schematic diagram illustrating nanocellulose surface functionalization modification. PEG, poly(ethylene glycol); PEO, poly(ethylene oxide); PLA, poly(lactic acid); PAA, poly(acrylic acid); PNiPAAm, poly(N-isopropylacrylamide); PDMAEMA, poly(N,N-dimethylaminoethyl methacrylate). Adapted from Ref. [99].

based on the "grafting from" approach with a radical polymerization involving single-electron transfer-living radical polymerization (SET-LP), ring opening polymerization (ROP), and atom transfer radical polymerization (ATRP) (as indicated with yellow arrows in **Figure 9**). The enhancement of NCC-polymer matrix interaction is predicted to improve the stress transfer from the matrix to the dispersed phase and hence enhances the capability of load bearing material. Besides, the chemical modification of NCC can be dispersed in the low polarity of organic solvent and mixed with a polymer matrix solution or directly introduced into the polymer melt after drying. Nevertheless, two effects ascend from this process: (1) allow the improvement of dispersion of modified NCC in the polymer matrix and (2) limit the interaction between NNC and matrix through hydrogen bonding which is the basis of the outstanding mechanical properties of nanocellulose-based nanocomposites.

6. Applications of nanocrystalline cellulose from agro-waste fiber and forest by-products

The incorporation of nanocrystalline cellulose in biopolymers for the nanocomposite production provides huge advantages with superior performance which would extend their applications in various applications. This is due to their outstanding thermal and mechanical properties. NCC also can reduce the water vapor permeability of the composites due to its high gas permeability [26]. Besides that, NCC can be used to stabilize the encapsulated bioactive compounds in biopolymers for allowing better control in food applications which can improve the food quality, extend the shelf-life of food, and serve as active substance carriers such as antifungal, antioxidant compounds, antimicrobial, and insecticide.

The utilization of natural cellulose-based materials continues today as verified by the various industry players from forest product to make pulp and paper to the advanced technology used in biomedical applications. These uses have been reported extensively as summarized in **Table 4**. NCC can be used as a drug delivery

Polymer component	Manufacturing technique	Applications	References
Cellulose esterified with lauroyl chloride	Solution casting and thermopressing	Interface melting	[101]
Ethyl acrylate; methyl-methacrylate	Solution mixing	Drug carrier	[100]
Ethylene-co-vinyl acetate rubber	Solution mixing and vulcanization	Transparent, rubbery materials	[102]
Maleic-anhydride grafted PLA	Electrospinning	Bone tissue engineering	[103]
Methylcellulose	Hydrogel by aqueous dispersion	Thermoreversible and tunable nanocellulose-based hydrogels	[104]
PC	Masterbatch melt extrusion process	Optical devices	[105]
PC-based polyurethane blend	Solution casting	Smart actuators and sensors	[106]
Plasticized PLA	Twin-screw extruder	Film blowing, packaging	[107]
Plasticized starch	Solution casting	Transparent materials	[108]

Polymer component	Manufacturing technique	Applications	References
PU	Solution casting	High temperature biomedical devices	[109]
PVA	Solution casting	Stretchable photonic devices	[110]
PVA	Solution casting	Wound diagnosis/biosensor scaffolds	[111]
PVA	Solution casting	Conductive materials	[112]
Starch	Blending, solution casting	Air permeable, resistant, surface-sized paper, food packaging	[113, 114]
Starch	Solution casting	Food packaging	[60]
Cassava starch	Solution casting	Food packaging	[62]
Sugar palm starch	Solution casting	Food packaging	[115]
Wheat starch	Solution casting	Food packaging	[87]
Tuber native potato	Solution casting	Packaging	[116]
Cereal corn	Solution casting	Packaging	[116]
Legume pea	Solution casting	Packaging	[116]
Waterborne acrylate	Solution mixing	Corrosion protection	[79]
Wheat straw hemicelluloses	Solution casting	Packaging	[96]
PVA	Solution casting	Food packaging	[83]
Chitosan	Solution casting	Food coating/packaging	[70]

Table 4.
Polymer component reinforced NCCs and its manufacturing technique and applications.

excipient; Burt et al. [100] investigated the capability of pure NCC to bind water-soluble antibiotics (tetracycline and doxorubicin) and the potential of cationic NCC to bind non-ionized hydrophobic anticancer agents (docetaxel, paclitaxel, and etoposide). Moreover, besides direct use as drug delivery excipient, NCC can also be used as co-stabilizer to improve the physicochemical and flow properties of polymeric excipients. Acrylic beads prepared via emulsion polymerization using NCC as co-stabilizer were proven to be a suitable excipient.

Table 5 shows several nanocelluloses, NFCs, and NCCs that have been used as reinforcement fillers in polymer matrices. The polymer matrices used are from both synthetic and natural polymers. **Table 6** shows examples of NCCs used as fillers in polymeric matrices.

Source	Filler	Polymer matrix	Ref.
Sugar palm	NCC	Sugar palm starch	[117]
Sugar palm	NFC	Sugar palm starch	[115]
Acacia mangium	NCC	PVA	[56]
Bacteria	NCC	CAB (0–10 wt% filler)	[58]
Cotton	NCC	PVA (0–12 wt% filler)	[118]
Flax	NFC/ NCC	PVA (10 wt% filler), waterborne polyurethanes (0–30 wt% filler)	[119, 120]
Hemp	NFC	PVA (10 wt% filler)	[120]
Kraft pulp	NCC	Waterborne acrylate	[79]

Source	Filler	Polymer matrix	Ref.
MCC	NCC	PLA (5 wt% filler)	[121]
Potato pulp	NFC	Starch/glycerol (0–40 wt% filler)	[122]
Ramie	NFC	Unsaturated polyester resin	[85]
Ramie	NCC	Starch/glycerol (0–40 wt% filler)	[87]
Rutabaga	NFC	PVA (10 wt% filler)	[120]
Soy hulls	NFC	No attempts were made with composites	[29]
Sugar beet	NFC/NCC	Styrene/butyl acrylate (6 wt% filler)	[123]
Tunicate	NCC	Styrene/butyl acrylate (6 wt% filler), starch/sorbitol (25 wt% filler), waterborne epoxy (0.5–5 wt% filler)	[94, 124–126]
Water hyacinth fiber	NCC	Yam bean starch	[48]
Water hyacinth fiber	NFC	Yam bean starch	[127]
Wheat straw	NFC	No attempts were made with composites	[29]
Wheat straw	NCC	Wheat straw hemicelluloses	[96]
Wood pulp	NFC/NCC	PVA (10 wt% filler), PLA (5 wt% filler)	[120, 128]
Cassava bagasse	NCC	Cassava starch	[62]
Ramie	NCC	Wheat starch	[87]
Phormium tenax (harakeke) fiber	NCC	PVA	[83]
Flax fiber	NCC	PVA	[83]
Potato peel fiber	NCC	Starch	[84]

Table 5.
Different nanocellulose sources of reinforcement fillers in polymer matrices.

Polymer	References
Cellulose acetate butyrate	[58, 129]
Cellulose	[130]
Chitosan	[131–133]
Poly(acrylic) acid, PAA	[134]
Poly-(allylmethylamine hydrochloride), PAH	[135]
Poly-(dimethyldiallylammonium chloride), PDDA	[136]
Poly(ethylene-co-vinyl acetate), EVA	[137]
Poly(hydroxyalkanoate), PHA	[133, 138]
Poly(hydroxyoctanoate), PHO	[139]
Poly(lactic acid), PLA	[118, 121, 140–144]
Poly(methyl-methacrylate), PMMA	[145, 146]
Poly(oxyethylene), PEO	[147, 148]

Polymer	References
Poly(styrene-co-butyl acrylate)	[94, 149, 150]
Poly(vinyl alcohol) (PVA)	[56, 83, 151]
Poly(vinyl alcohol), PVOH	[67, 152–154]
Polycaprolactone, PCL	[155–157]
Polypropylene, PP	[158, 159]
Polystyrene	[160]
Polysulfone	[161]
Polyurethane, PU	[162–164]
Polyvinyl chloride, PVC	[165–167]
Regenerated cellulose	[168, 169]
Soy protein	[170]
Starch-based polymers	[60, 62, 84, 152, 171–173]
Waterborne acrylate	[79]
Xylan	[174–176]
Hemicellulose	[96]

Table 6.
NCC used as filler in polymeric matrices.

7. Conclusion

Agro-waste is an unavoidable by-product that arises from various agricultural and agro-forest activities' operation. However, different kinds of agro-product industries, change of lifestyle, and population growth are assumed to be within the main factors that increase the rate of waste generation globally and locally. Therefore, proper waste management selections are very important based on the types of wastes and cost-effective factors in order to reduce the damage to the ecosystem. One of the alternatives to reduce agro-waste disposal is converting it to high-end value products such as nanocrystalline cellulose. In the present work, an overview of the production, processes, modification, and application of nanocrystalline cellulose from different agricultural wastes was proposed and leads to the following main concluding remarks: (1) it is important to select the proper raw material of agro-waste fiber, due to a broad variety of structure and chemical composition and its pretreatment process before the extraction process of nanocellulose begin; (2) the surface charge and morphology of nanocrystalline cellulose are affected by the production conditions such as hydrolysis time, temperature, and the acid-to-fiber ratio; and (3) nanocrystalline cellulose can be used in various applications including in hydrophobic polymer after some modification is made. The utilization of several lignocellulosic wastes from agricultural and forest by-product activities becomes the best proposal regarding cost/energy savings and economic development. The agricultural residue is available worldwide, abundant, cheap, and an unexploited source of cellulose that could be used as large-scale production of nanocellulose products.

Acknowledgements

The authors would like to thank Universiti Putra Malaysia for the financial support through the Graduate Research Fellowship (GRF) scholarship, Universiti

Putra Malaysia Grant scheme Hi-CoE (6369107), FRGS/1/2017/TK05/UPM/01/1 (5540048) and iRMC UNITEN (RJO10436494). The authors are grateful to Dr. Muhammed Lamin Sanyang for guidance throughout the experiment. The authors also thank Dr. Rushdan Ibrahim for his advice and fruitful discussion.

Author details

R.A. Ilyas[1,2*], S.M. Sapuan[1,2], R. Ibrahim[3], M.S.N. Atikah[4], A. Atiqah[5], M.N.M. Ansari[5] and M.N.F. Norrrahim[6]

1 Laboratory of Biocomposite Technology, Institute of Tropical Forestry and Forest Products, Universiti Putra Malaysia, Serdang, Selangor, Malaysia

2 Department of Mechanical and Manufacturing Engineering, Advanced Engineering Materials and Composites Research Centre, Universiti Putra Malaysia, Serdang, Selangor, Malaysia

3 Pulp and Paper Branch, Forest Research Institute Malaysia, Kepong, Selangor, Malaysia

4 Department of Chemical and Environmental Engineering, Universiti Putra Malaysia, Serdang, Selangor, Malaysia

5 Institute of Power Engineering, Universiti Tenaga Nasional, Kajang, Selangor, Malaysia

6 Research Center for Chemical Defence, Universiti Pertahanan Nasional Malaysia, Kuala Lumpur, Malaysia

*Address all correspondence to: ahmadilyasrushdan@yahoo.com

IntechOpen

© 2019 The Author(s). Licensee IntechOpen. This chapter is distributed under the terms of the Creative Commons Attribution License (http://creativecommons.org/licenses/by/3.0), which permits unrestricted use, distribution, and reproduction in any medium, provided the original work is properly cited. (cc) BY

References

[1] Hazrati KZ, Sapuan SM, Ilyas RA. Biobased food packaging using natural fibre: A review. In: Prosiding Seminar Enau Kebangsaan 2019, Bahau, Negeri Sembilan, Malaysia: Institute of Tropical Forest and Forest Products. Universiti Putra Malaysia: INTROP; 2019. pp. 140-142

[2] Abral H, Basri A, Muhammad F, Fernando Y, Hafizulhaq F, Mahardika M, et al. A simple method for improving the properties of the sago starch films prepared by using ultrasonication treatment. Food Hydrocolloids. 2019;**93**: 276-283. DOI: 10.1016/j.foodhyd.2019.02.012

[3] Huzaifah MRM, Sapuan SM, Leman Z, Ishak MR, Ilyas RA. Effect of soil burial on water absorption of sugar palm fibre reinforced vinyl ester composites. In: 6th Postgraduate Seminar on Natural Fiber Reinforced Polymer Composites 2018. Selangor: Serdang; 2018. pp. 52-54

[4] Brinchi L, Cotana F, Fortunati E, Kenny JM. Production of nanocrystalline cellulose from lignocellulosic biomass: Technology and applications. Carbohydrate Polymers. 2013;**94**:154-169. DOI: 10.1016/j.carbpol.2013.01.033

[5] Norizan MN, Abdan K, Ilyas RA. Effect of water absorption on treated sugar palm yarn fibre/glass fibre hybrid composites. In: Prosiding Seminar Enau Kebangsaan 2019, Bahau, Negeri Sembilan, Malaysia: Institute of Tropical Forest and Forest Products. Universiti Putra Malaysia: INTROP; 2019. pp. 78-81

[6] Ilyas RA, Sapuan SM, Ishak MR. Isolation and characterization of nanocrystalline cellulose from sugar palm fibres (*Arenga pinnata*). Carbohydrate Polymers. 2018;**181**: 1038-1051. DOI: 10.1016/j.carbpol.2017.11.045

[7] Ilyas RA, Sapuan SM, Ishak MR, Zainudin ES, Atikah MSN. Characterization of sugar palm nanocellulose and its potential for reinforcement with a starch-based composite. In: Sugar Palm Biofibers, Biopolymers, and Biocomposites. 1st ed. Boca Raton, FL: CRC Press/Taylor & Francis Group; 2018. pp. 189-220. DOI: 10.1201/9780429443923-10

[8] Sanyang ML, Ilyas RA, Sapuan SM, Jumaidin R. Sugar palm starch-based composites for packaging applications. In: Bionanocomposites for Packaging Applications. Cham: Springer International Publishing; 2018. pp. 125-147. DOI: 10.1007/978-3-319-67319-6_7

[9] Azammi AMN, Sapuan SM, Sultan MTH, Ishak MR, Radzi AM, Ilyas RA. Structure analysis for natural fiber composite for automotive component: A review. In: Prosiding Seminar Enau Kebangsaan 2019, Bahau, Negeri Sembilan, Malaysia: Institute of Tropical Forest and Forest Products. Universiti Putra Malaysia: INTROP; 2019. pp. 44-47

[10] Ilyas RA, Sapuan SM, Sanyang ML, Ishak MR, Zainudin ES. Nanocrystalline cellulose as reinforcement for polymeric matrix nanocomposites and its potential applications: A review. Current Analytical Chemistry. 2018;**14**:203-225. DOI: 10.2174/1573411013666171003155624

[11] Mazani N, Sapuan SM, Sanyang ML, Atiqah A, Ilyas RA. Design and fabrication of a shoe shelf from kenaf fiber reinforced unsaturated polyester composites. In: Lignocellulose for Future Bioeconomy. Elsevier; 2019. pp. 315-332. DOI: 10.1016/B978-0-12-816354-2.00017-7

[12] Ilyas RA, Sapuan SM, Ibrahim R, Abral H, Ishak MR, Zainudin ES, et al. Sugar palm (*Arenga pinnata (Wurmb.) Merr*) cellulosic fibre hierarchy: A comprehensive approach from macro to nano scale. Journal of Materials Research and Technology. 2019. DOI: 10.1016/j.jmrt.2019.04.011

[13] Liu C, Li B, Du H, Lv D, Zhang Y, Yu G, et al. Properties of nanocellulose isolated from corncob residue using sulfuric acid, formic acid, oxidative and mechanical methods. Carbohydrate Polymers. 2016;**151**:716-724. DOI: 10.1016/j.carbpol.2016.06.025

[14] Julie Chandra CS, George N, Narayanankutty SK. Isolation and characterization of cellulose nanofibrils from arecanut husk fibre. Carbohydrate Polymers. 2016;**142**:158-166. DOI: 10.1016/j.carbpol.2016.01.015

[15] Ilyas RA, Sapuan SM, Ishak MR, Zainudin ES. Sugar palm nanocrystalline cellulose reinforced sugar palm starch composite: Degradation and water-barrier properties. IOP Conference Series: Materials Science and Engineering. 2018;**368**. DOI: 10.1088/1757-899X/368/1/012006

[16] Sapuan SM, Ishak MR, Leman Z, Ilyas RA, Huzaifah MRM. Development of products from sugar palm trees (*Arenga pinnata Wurb. Merr*): A community project. INTROPica. 2017: 12-13

[17] Jumaidin R, Sapuan SM, Ilyas RA. Physio-mechanical properties of thermoplastic starch composites: A review. In: Prosiding Seminar Enau Kebangsaan 2019, Bahau, Negeri Sembilan, Malaysia: Institute of Tropical Forest and Forest Products. Universiti Putra Malaysia: INTROP; 2019. pp. 104-108

[18] Hazrol MD, Sapuan SM, Ilyas RA. Electrical and surface resistivity of polymer composites: A review. In: 6th Postgraduate Seminar on Natural Fiber Reinforced Polymer Composites 2018; Serdang, Selangor. 2018. pp. 44-47

[19] Nazrin A, Sapuan SM, Ilyas RA. Thermoplastic starch blended poly (lactic) acid for food packaging application: Mechanical properties. In: 6th Postgraduate Seminar on Natural Fiber Reinforced Polymer Composites 2018, Serdang. 2018. pp. 79-84

[20] Sapuan SM, Ilyas RA. Sugar palm: Fibers, biopolymers and biocomposites. INTROPica. 2017:5-7

[21] Halimatul MJ, Sapuan SM, Jawaid M, Ishak MR, Ilyas RA. Effect of sago starch and plasticizer content on the properties of thermoplastic films: Mechanical testing and cyclic soaking-drying. Polimery. 2019;**64**:422-431. DOI: 10.14314/polimery.2019.6.5

[22] Ilyas RA, Sapuan SM, Norizan MN, Atikah MSN, Huzaifah MRM, Radzi AM, et al. Potential of natural fibre composites for transport industry: A review. In: Prosiding Seminar Enau Kebangsaan 2019, Bahau, Negeri Sembilan, Malaysia: Institute of Tropical Forest and Forest Products. Universiti Putra Malaysia: INTROP; 2019. pp. 2-11

[23] Solid Bioenergy—An Asset to EU Forests? European Bioenergy Day 2018. n.d. Available from: http://www.europeanbioenergyday.eu/solid-bioenergy-in-questions-an-asset-to-eu-forests/ [Accessed: April 29, 2019]

[24] Ilyas RA, Sapuan SM, Ishak MR, Zainudin ES, Atikah MSN. Nanocellulose reinforced starch polymer composites: A review of preparation, properties and application. In: Proceeding: 5th International Conference on Applied Sciences and Engineering (ICASEA, 2018), Copthorne Hotel; Cameron Highlands, Malaysia. 2018. pp. 325-341

[25] Ilyas RA, Sapuan SM, Ishak MR, Zainudin ES. Sugar palm nanofibrillated cellulose (*Arenga pinnata (Wurmb.) Merr*): Effect of cycles on their yield, physic-chemical, morphological and thermal behavior. International Journal of Biological Macromolecules. 2019;**123**: 379-388. DOI: 10.1016/j.ijbiomac. 2018.11.124

[26] Ilyas RA, Sapuan SM, Ishak MR, Zainudin ES, Atikah MSN, Huzaifah MRM. Water barrier properties of biodegradable films reinforced with nanocellulose for food packaging application: A review. In: 6th Postgraduate Seminar on Natural Fiber Reinforced Polymer Composites 2018; Serdang, Selangor. 2018. pp. 55-59

[27] Ilyas RA, Sapuan SM, Ishak MR, Zainudin ES. Effect of delignification on the physical, thermal, chemical, and structural properties of sugar palm fibre. BioResources. 2017;**12**:8734-8754. DOI: 10.15376/biores.12.4.8734-8754

[28] Sanyang ML, Sapuan SM, Jawaid M, Ishak MR, Sahari J. Recent developments in sugar palm (*Arenga pinnata*) based biocomposites and their potential industrial applications: A review. Renewable and Sustainable Energy Reviews. 2016;**54**:533-549. DOI: 10.1016/j.rser.2015.10.037

[29] Alemdar A, Sain M. Isolation and characterization of nanofibers from agricultural residues—Wheat straw and soy hulls. Bioresource Technology. 2008;**99**:1664-1671. DOI: 10.1016/j. biortech.2007.04.029

[30] Chirayil CJ, Joy J, Mathew L, Mozetic M, Koetz J, Thomas S. Isolation and characterization of cellulose nanofibrils from *Helicteres isora* plant. Industrial Crops and Products. 2014;**59**: 27-34. DOI: 10.1016/j.indcrop. 2014.04.020

[31] Cherian BM, Leão AL, de Souza SF, Thomas S, Pothan LA, Kottaisamy M. Isolation of nanocellulose from pineapple leaf fibres by steam explosion. Carbohydrate Polymers. 2010;**81**: 720-725. DOI: 10.1016/j.carbpol. 2010.03.046

[32] Syafri E, Kasim A, Abral H, Asben A. Cellulose nanofibers isolation and characterization from ramie using a chemical-ultrasonic treatment. Journal of Natural Fibers. 2018;**00**:1-11. DOI: 10.1080/15440478.2018.1455073

[33] Megashah LN, Ariffin H, Zakaria MR, Hassan MA. Properties of cellulose extract from different types of oil palm biomass. IOP Conference Series: Materials Science and Engineering. 2018;**368**. DOI: 10.1088/ 1757-899X/368/1/012049

[34] Jonoobi M, Khazaeian A, Tahir PM, Azry SS, Oksman K. Characteristics of cellulose nanofibers isolated from rubberwood and empty fruit bunches of oil palm using chemo-mechanical process. Cellulose. 2011;**18**:1085-1095. DOI: 10.1007/s10570-011-9546-7

[35] Corrêa AC, de Morais Teixeira E, Pessan LA, Mattoso LHC. Cellulose nanofibers from curaua fibers. Cellulose. 2010;**17**:1183-1192. DOI: 10.1007/s10570-010-9453-3

[36] Tibolla H, Pelissari FM, Menegalli FC. Cellulose nanofibers produced from banana peel by chemical and enzymatic treatment. LWT—Food Science and Technology. 2014;**59**: 1311-1318. DOI: 10.1016/j. lwt.2014.04.011

[37] De ME, Jessika T, Bruna K, Teodoro R, Carolina A, Manoel J, et al. Sugarcane bagasse whiskers: Extraction and characterizations. Industrial Crops and Products. 2011;**33**:66. DOI: 10.1016/ j.indcrop.2010.08.009

[38] Jonoobi M, Harun J, Shakeri A, Misra M, Oksmand K. Chemical composition, crystallinity, and thermal degradation of bleached and unbleached kenaf bast (*Hibiscus cannabinus*) pulp

and nanofibers. BioResources. 2009;**4**: 626-639. DOI: 10.15376/biores. 4.2.626-639

[39] Bendahou A, Habibi Y, Kaddami H, Dufresne A. Physico-chemical characterization of palm from *Phoenix dactylifera*-L, preparation of cellulose whiskers and natural rubber-based nanocomposites. Journal of Biobased Materials and Bioenergy. 2009;**3**:81-90. DOI: 10.1166/jbmb.2009.1011

[40] Chan CH, Chia CH, Zakaria S, Ahmad I, Dufresne A. Production and characterisation of cellulose and nano-crystalline cellulose from kenaf core wood. BioResources. 2013;**8**:785-794. DOI: 10.15376/biores.8.1.785-794

[41] Abral H, Dalimunthe MH, Hartono J, Efendi RP, Asrofi M, Sugiarti E, et al. Characterization of tapioca starch biopolymer composites reinforced with micro scale water hyacinth fibers. Starch/Staerke. 2018;**70**: 1-8. DOI: 10.1002/star.201700287

[42] Alemdar A, Sain M. Biocomposites from wheat straw nanofibers: Morphology, thermal and mechanical properties. Composites Science and Technology. 2008;**68**:557-565. DOI: 10.1016/j.compscitech.2007.05.044

[43] Li M, Wang LJ, Li D, Cheng YL, Adhikari B. Preparation and characterization of cellulose nanofibers from de-pectinated sugar beet pulp. Carbohydrate Polymers. 2014;**102**: 136-143. DOI: 10.1016/j.carbpol. 2013.11.021

[44] Sheltami RM, Abdullah I, Ahmad I, Dufresne A, Kargarzadeh H. Extraction of cellulose nanocrystals from mengkuang leaves (*Pandanus tectorius*). Carbohydrate Polymers. 2012;**88**: 772-779. DOI: 10.1016/j.carbpol. 2012.01.062

[45] Ilyas RA, Sapuan SM, Sanyang ML, Ishak MR. Nanocrystalline cellulose reinforced starch-based nanocomposite: A review. In: 5th Postgraduate Seminar on Natural Fiber Composites, Serdang, Selangor: Universiti Putra Malaysia. 2016. pp. 82-87

[46] Bagheri S, Julkapli NM, Mansouri N. Nanocrystalline cellulose: Green, multifunctional and sustainable nanomaterials. In: Handbook of Composites from Renewable Materials. Vol. 1–8. 2017. pp. 523-555. DOI: 10.1002/9781119441632.ch142

[47] CelluForce. Cellulose Nanocrystals 2016. Available from: https://www.celluforce.com/en/products/cellulose-nanocrystals/ [Accessed: May 1, 2019]

[48] Asrofi M, Abral H, Kasim A, Pratoto A, Mahardika M, Hafizulhaq F. Characterization of the sonicated yam bean starch bionanocomposites reinforced by nanocellulose water hyacinth fiber (Whf): The effect of various fiber loading. Journal of Engineering Science and Technology. 2018;**13**:2700-2715

[49] Lee S-Y, Mohan DJ, Kang I-A, Doh G-H, Lee S, Han SO. Nanocellulose reinforced PVA composite films: Effects of acid treatment and filler loading. Fibers and Polymers. 2009; **10**:77-82. DOI: 10.1007/s12221-009-0077-x

[50] Camarero Espinosa S, Kuhnt T, Foster EJ, Weder C. Isolation of thermally stable cellulose nanocrystals by phosphoric acid hydrolysis. Biomacromolecules. 2013;**14**:1223-1230. DOI: 10.1021/bm400219u

[51] Beck-Candanedo S, Roman M, Gray DG. Effect of reaction conditions on the properties and behavior of wood cellulose nanocrystal suspensions. Biomacromolecules. 2005;**6**:1048-1054

[52] Roman M, Winter WT. Effect of sulfate groups from sulfuric acid hydrolysis on the thermal degradation behavior of bacterial cellulose.

Biomacromolecules. 2004;**5**:1671-1677. DOI: 10.1021/bm034519+

[53] Xiang Q, Lee YY, Pettersson PO, Torget RW. Heterogeneous aspects of acid hydrolysis of α-cellulose. In: Biotechnology for Fuels and Chemicals. Totowa, NJ: Humana Press; 2003. pp. 505-514. DOI: 10.1007/978-1-4612-0057-4_42

[54] Azizi Samir MAS, Alloin F, Dufresne A. Review of recent research into cellulosic whiskers, their properties and their application in nanocomposite field. Biomacromolecules. 2005;**6**: 612-626. DOI: 10.1021/bm0493685

[55] Bondeson D, Mathew A, Oksman K. Optimization of the isolation of nanocrystals from microcrystalline cellulose by acid hydrolysis. Cellulose. 2006;**13**:171-180. DOI: 10.1007/s10570-006-9061-4

[56] Jasmani L, Adnan S. Preparation and characterization of nanocrystalline cellulose from *Acacia mangium* and its reinforcement potential. Carbohydrate Polymers. 2017;**161**:166-171. DOI: 10.1016/j.carbpol.2016.12.061

[57] Imai T, Putaux J, Sugiyama J. Geometric phase analysis of lattice images from algal cellulose microfibrils. Polymer. 2003;**44**:1871-1879. DOI: 10.1016/S0032-3861(02)00861-3

[58] Grunert M, Winter WT. Nanocomposites of cellulose acetate butyrate reinforced with cellulose nanocrystals. Journal of Polymers and the Environment. 2002;**10**:27-30

[59] Brito BSL, Pereira FV, Putaux J-L, Jean B. Preparation, morphology and structure of cellulose nanocrystals from bamboo fibers. Cellulose. 2012;**19**: 1527-1536. DOI: 10.1007/s10570-012-9738-9

[60] Liu D, Zhong T, Chang PR, Li K, Wu Q. Starch composites reinforced by bamboo cellulosic crystals. Bioresource Technology. 2010;**101**:2529-2536. DOI: 10.1016/j.biortech.2009.11.058

[61] Faradilla RHF, Lee G, Arns JY, Roberts J, Martens P, Stenzel MH, et al. Characteristics of a free-standing film from banana pseudostem nanocellulose generated from TEMPO-mediated oxidation. Carbohydrate Polymers. 2017;**174**:1156-1163. DOI: 10.1016/j.carbpol.2017.07.025

[62] Teixeira E d M, Pasquini D, Curvelo AASS, Corradini E, Belgacem MN, Dufresne A. Cassava bagasse cellulose nanofibrils reinforced thermoplastic cassava starch. Carbohydrate Polymers. 2009;**78**: 422-431. DOI: 10.1016/j.carbpol.2009.04.034

[63] Rosa MFM, Medeiros ES, Malmonge JAJ, Gregorski KS, Wood DF, Mattoso LHC, et al. Cellulose nanowhiskers from coconut husk fibers: Effect of preparation conditions on their thermal and morphological behavior. Carbohydrate Polymers. 2010;**81**:83-92. DOI: 10.1016/j.carbpol.2010.01.059

[64] de Morais Teixeira E, Corrêa AC, Manzoli A, de Lima Leite F, de Ribeiro Oliveira C, Mattoso LHC. Cellulose nanofibers from white and naturally colored cotton fibers. Cellulose. 2010;**17**: 595-606. DOI: 10.1007/s10570-010-9403-0

[65] Morandi G, Heath L, Thielemans W. Cellulose nano-crystals grafted with polystyrene chains through surface-initiated atom transfer radical polymerization (SI-ATRP). Langmuir. 2009;**25**:8280-8286

[66] Braun B, Dorgan JR, Chandler JP. Cellulosic nanowhiskers. Theory and application of light scattering from polydisperse spheroids in the Rayleigh–Gans–Debye regime. Biomacromolecules. 2008;**9**:1255-1263. DOI: 10.1021/bm7013137

[67] Paralikar SA, Simonsen J, Lombardi J. Poly(vinyl alcohol)/cellulose nanocrystal barrier membranes. Journal of Membrane Science. 2008;**320**:248-258. DOI: 10.1016/j.memsci.2008.04.009

[68] Morais JPS, Rosa MDF, De Souza Filho MM, Nascimento LD, Do Nascimento DM, Cassales AR. Extraction and characterization of nanocellulose structures from raw cotton linter. Carbohydrate Polymers. 2013;**91**:229-235. DOI: 10.1016/j.carbpol.2012.08.010.

[69] Soni B, Hassan EB, Mahmoud B. Chemical isolation and characterization of different cellulose nanofibers from cotton stalks. Carbohydrate Polymers. 2015;**134**:581-589. DOI: 10.1016/j.carbpol.2015.08.031

[70] Pereda M, Dufresne A, Aranguren MI, Marcovich NE. Polyelectrolyte films based on chitosan/olive oil and reinforced with cellulose nanocrystals. Carbohydrate Polymers. 2014;**101**:1018-1026. DOI: 10.1016/j.carbpol.2013.10.046

[71] Tonoli GHDHD, Teixeira EMM, Corrêa ACC, Marconcini JMM, Caixeta LAA, Pereira-Da-Silva MAA, et al. Cellulose micro/nanofibres from eucalyptus kraft pulp: Preparation and properties. Carbohydrate Polymers. 2012;**89**:80-88. DOI: 10.1016/j.carbpol.2012.02.052

[72] Pandey JK, Kim C, Chu W, Lee CS, Jang D-Y, Ahn S. Evaluation of morphological architecture of cellulose chains in grass during conversion from macro to nano dimensions. E-Polymers. 2009;**9**:1-15. DOI: 10.1515/epoly.2009.9.1.1221

[73] Benini KCCC, Voorwald HJC, Cioffi MOH, Rezende MC, Arantes V. Preparation of nanocellulose from Imperata brasiliensis grass using Taguchi method. Carbohydrate Polymers. 2018;**192**:337-346. DOI: 10.1016/j.carbpol.2018.03.055

[74] Bano S, Negi YS. Studies on cellulose nanocrystals isolated from groundnut shells. Carbohydrate Polymers. 2017;**157**:1041-1049. DOI: 10.1016/j.carbpol.2016.10.069

[75] Sonia A, Priya Dasan K. Chemical, morphology and thermal evaluation of cellulose microfibers obtained from *Hibiscus sabdariffa*. Carbohydrate Polymers. 2013;**92**:668-674. DOI: 10.1016/j.carbpol.2012.09.015

[76] Jiang Y, Zhou J, Zhang Q, Zhao G, Heng L, Chen D, et al. Preparation of cellulose nanocrystals from *Humulus japonicus* stem and the influence of high temperature pretreatment. Carbohydrate Polymers. 2017;**164**:284-293. DOI: 10.1016/j.carbpol.2017.02.021

[77] Oksman K, Etang JA, Mathew AP, Jonoobi M. Cellulose nanowhiskers separated from a bio-residue from wood bioethanol production. Biomass and Bioenergy. 2010;**35**:146-152. DOI: 10.1016/j.biombioe.2010.08.021

[78] Herrera MA, Mathew AP, Oksman K. Comparison of cellulose nanowhiskers extracted from industrial bio-residue and commercial microcrystalline cellulose. Materials Letters. 2012;**71**:28-31. DOI: 10.1016/j.matlet.2011.12.011

[79] He Y, Boluk Y, Pan J, Ahniyaz A, Deltin T, Claesson PM. Corrosion protective properties of cellulose nanocrystals reinforced waterborne acrylate-based composite coating. Corrosion Science. 2019. DOI: 10.1016/j.corsci.2019.04.038

[80] Li R, Fei J, Cai Y, Li Y, Feng J, Yao J. Cellulose whiskers extracted from mulberry: A novel biomass production. Carbohydrate Polymers. 2009;**76**:94-99. DOI: 10.1016/j.carbpol.2008.09.034

[81] Lamaming J, Hashim R, Sulaiman O, Leh CP, Sugimoto T, Nordin NA. Cellulose nanocrystals isolated from oil palm trunk. Carbohydrate Polymers. 2015;**127**:202-208. DOI: 10.1016/j.carbpol.2015.03.043

[82] Haafiz MKM, Hassan A, Zakaria Z, Inuwa IM. Isolation and characterization of cellulose nanowhiskers from oil palm biomass microcrystalline cellulose. Carbohydrate Polymers. 2014;**103**:119-125. DOI: 10.1016/j.carbpol.2013.11.055

[83] Fortunati E, Puglia D, Luzi F, Santulli C, Kenny JM, Torre L. Binary PVA bio-nanocomposites containing cellulose nanocrystals extracted from different natural sources: Part I. Carbohydrate Polymers. 2013;**97**: 825-836. DOI: 10.1016/j.carbpol.2013.03.075

[84] Chen D, Lawton D, Thompson MR, Liu Q. Biocomposites reinforced with cellulose nanocrystals derived from potato peel waste. Carbohydrate Polymers. 2012;**90**:709-716. DOI: 10.1016/j.carbpol.2012.06.002

[85] Wahono S, Irwan A, Syafri E, Asrofi M. Preparation and characterization of ramie cellulose nanofibers/$CaCO_3$ unsaturated polyester resin composites. ARPN Journal of Engineering and Applied Sciences. 2018;**13**:746-751. DOI: 10.1039/c7nr02736b

[86] Habibi Y, Vignon MR. Optimization of cellouronic acid synthesis by TEMPO-mediated oxidation of cellulose III from sugar beet pulp. Cellulose. 2008;**15**:177-185. DOI: 10.1007/s10570-007-9179-z

[87] Lu Y, Weng L, Cao X. Morphological, thermal and mechanical properties of ramie crystallites—Reinforced plasticized starch biocomposites. Carbohydrate Polymers. 2006;**63**:198-204. DOI: 10.1016/j.carbpol.2005.08.027

[88] Lu P, Hsieh Y. Preparation and characterization of cellulose nanocrystals from rice straw. Carbohydrate Polymers. 2012;**87**: 564-573. DOI: 10.1016/j.carbpol.2011.08.022

[89] Purkait BS, Ray D, Sengupta S, Kar T, Mohanty A, Misra M. Isolation of cellulose nanoparticles from sesame husk. Industrial & Engineering Chemistry Research. 2011;**50**:871-876. DOI: 10.1021/ie101797d

[90] Morán JI, Alvarez VA, Cyras VP, Vázquez A. Extraction of cellulose and preparation of nanocellulose from sisal fibers. Cellulose. 2008;**15**:149-159. DOI: 10.1007/s10570-007-9145-9

[91] Flauzino Neto WP, Silvério HA, Dantas NO, Pasquini D. Extraction and characterization of cellulose nanocrystals from agro-industrial residue—Soy hulls. Industrial Crops and Products. 2013;**42**:480-488. DOI: 10.1016/j.indcrop.2012.06.041

[92] Sumaiyah L, Wirjosentono B, Karsono, Nasution MP. Preparation and characterization of nanocrystalline cellulose from sugar palm bunch. Interantional Journal of PharmTech Research. 2014;**6**:814-820

[93] Naduparambath S, Jinitha TV, Shaniba V, Sreejith MP, Balan AK, Purushothaman E. Isolation and characterisation of cellulose nanocrystals from sago seed shells. Carbohydrate Polymers. 2018;**180**: 13-20. DOI: 10.1016/j.carbpol.2017.09.088.

[94] Favier V, Chanzy H, Cavaille JY. Polymer nanocomposites reinforced by cellulose whiskers. Macromolecules. 1995;**28**:6365-6367. DOI: 10.1021/ma00122a053

[95] Salajková M, Berglund LA, Zhou Q. Hydrophobic cellulose nanocrystals modified with quaternary ammonium salts. Journal of Materials Chemistry. 2012;**22**:19798. DOI: 10.1039/c2jm34355j

[96] Pereira PHF, Waldron KW, Wilson DR, Cunha AP, de Brito ES, Rodrigues THS, et al. Wheat straw hemicelluloses added with cellulose nanocrystals and citric acid. Effect on film physical properties. Carbohydrate Polymers. 2017;**164**:317-324. DOI: 10.1016/j.carbpol.2017.02.019

[97] Revol JF. On the cross-sectional shape of cellulose crystallites in *Valonia ventricosa*. Carbohydrate Polymers. 1982;**2**:123-134. DOI: 10.1016/0144-8617(82)90058-3

[98] Postek MT, Moon RJ, Rudie AW, Bilodeau MA. Production and Applications of Cellulose Nanomaterials. Peachtree Corners: Tappi Press; 2013

[99] Dufresne A. Nanocellulose: A new ageless bionanomaterial. Materials Today. 2013;**16**:220-227. DOI: 10.1016/j.mattod.2013.06.004

[100] Letchford J, Wasserman B, Ye HW, Burt H. The use of nanocrystalline cellulose for the binding and controlled release of drugs. International Journal of Nanomedicine. 2011:321. DOI: 10.2147/IJN.S16749

[101] Timhadjelt L, Serier A, Belgacem MN, Bras J. Elaboration of cellulose based nanobiocomposite: Effect of cellulose nanocrystals surface treatment and interface "melting.". Industrial Crops and Products. 2015;**72**: 7-15. DOI: 10.1016/j.indcrop.2015.02.040

[102] Ma P, Jiang L, Hoch M, Dong W, Chen M. Reinforcement of transparent ethylene-co-vinyl acetate rubber by nanocrystalline cellulose. European Polymer Journal. 2015;**66**:47-56. DOI: 10.1016/j.eurpolymj.2015.01.037

[103] Zhou C, Shi Q, Guo W, Terrell L, Qureshi AT, Hayes DJ, et al. Electrospun bio-Nanocomposite scaffolds for bone tissue engineering by cellulose Nanocrystals reinforcing maleic anhydride grafted PLA. ACS Applied Materials & Interfaces. 2013;**5**: 3847-3854. DOI: 10.1021/am4005072

[104] McKee JR, Hietala S, Seitsonen J, Laine J, Kontturi E, Ikkala O. Thermoresponsive nanocellulose hydrogels with tunable mechanical properties. ACS Macro Letters. 2014;**3**:266-270. DOI: 10.1021/mz400596g

[105] Mariano M, El Kissi N, Dufresne A. Melt processing of cellulose nanocrystal reinforced polycarbonate from a masterbatch process. European Polymer Journal. 2015;**69**:208-223 DOI: 10.1016/j.eurpolymj.2015.06.007

[106] Li Y, Chen H, Liu D, Wang W, Liu Y, Zhou S. pH-responsive shape memory poly(ethylene glycol)–poly(ε-caprolactone)-based polyurethane/cellulose nanocrystals nanocomposite. ACS Applied Materials & Interfaces. 2015;7:12988-12999. DOI: 10.1021/acsami.5b02940

[107] Herrera N, Salaberria AM, Mathew AP, Oksman K. Plasticized polylactic acid nanocomposite films with cellulose and chitin nanocrystals prepared using extrusion and compression molding with two cooling rates: Effects on mechanical, thermal and optical properties. Composites Part A: Applied Science and Manufacturing. 2016;**83**:89-97. DOI: 10.1016/j.compositesa.2015.05.024

[108] Nasseri R, Mohammadi N. Starch-based nanocomposites: A comparative performance study of cellulose whiskers and starch nanoparticles. Carbohydrate

Polymers. 2014;**106**:432-439. DOI: 10.1016/j.carbpol.2014.01.029

[109] Liu JC, Martin DJ, Moon RJ, Youngblood JP. Enhanced thermal stability of biomedical thermoplastic polyurethane with the addition of cellulose nanocrystals. Journal of Applied Polymer Science. 2015;**132**:1-8. DOI: 10.1002/app.41970

[110] Wang B, Walther A. Self-assembled, iridescent, crustacean-mimetic nanocomposites with tailored periodicity and layered cuticular structure. ACS Nano. 2015;**9**: 10637-10646. DOI: 10.1021/acsnano.5b05074

[111] Schyrr B, Pasche S, Voirin G, Weder C, Simon YC, Foster EJ. Biosensors based on porous cellulose nanocrystal–poly(vinyl alcohol) scaffolds. ACS Applied Materials & Interfaces. 2014;**6**:12674-12683. DOI: 10.1021/am502670u

[112] Montes S, Carrasco PM, Ruiz V, Cabañero G, Grande HJ, Labidi J, et al. Synergistic reinforcement of poly(vinyl alcohol) nanocomposites with cellulose nanocrystal-stabilized graphene. Composites Science and Technology. 2015;**117**:26-31. DOI: 10.1016/j.compscitech.2015.05.018

[113] Yang S, Tang Y, Wang J, Kong F, Zhang J. Surface treatment of cellulosic paper with starch-based composites reinforced with nanocrystalline cellulose. Industrial & Engineering Chemistry Research. 2014;**53**: 13980-13988. DOI: 10.1021/ie502125s

[114] Slavutsky AM, Bertuzzi MA. Water barrier properties of starch films reinforced with cellulose nanocrystals obtained from sugarcane bagasse. Carbohydrate Polymers. 2014;**110**:53-61. DOI: 10.1016/j.carbpol.2014.03.049

[115] Ilyas RA, Sapuan SM, Ishak MR, Zainudin ES. Water transport properties of bio-nanocomposites reinforced by sugar palm (*Arenga pinnata*) nanofibrillated cellulose. Journal of Advanced Research in Fluid Mechanics and Thermal Sciences Journal. 2018;**51**: 234-246

[116] Montero B, Rico M, Rodríguez-Llamazares S, Barral L, Bouza R. Effect of nanocellulose as a filler on biodegradable thermoplastic starch films from tuber, cereal and legume. Carbohydrate Polymers. 2017;**157**: 1094-1104. DOI: 10.1016/j.carbpol.2016.10.073

[117] Ilyas RA, Sapuan SM, Ishak MR, Zainudin ES. Development and characterization of sugar palm nanocrystalline cellulose reinforced sugar palm starch bionanocomposites. Carbohydrate Polymers. 2018;**202**: 186-202. DOI: 10.1016/j.carbpol.2018.09.002

[118] Roohani M, Habibi Y, Belgacem NM, Ebrahim G, Karimi AN, Dufresne A. Cellulose whiskers reinforced polyvinyl alcohol copolymers nanocomposites. European Polymer Journal. 2008;**44**:2489-2498. DOI: 10.1016/j.eurpolymj.2008.05.024

[119] Cao X, Xu C, Wang Y, Liu Y, Liu Y, Chen Y. New nanocomposite materials reinforced with cellulose nanocrystals in nitrile rubber. Polymer Testing. 2013;**32**: 819-826. DOI: 10.1016/j.polymertesting.2013.04.005

[120] Bhatnagar A. Processing of cellulose nanofiber-reinforced composites. Journal of Reinforced Plastics and Composites. 2005;**24**: 1259-1268. DOI: 10.1177/0731684405049864

[121] Kvien I, Tanem BS, Oksman K. Characterization of cellulose whiskers and their nanocomposites by atomic force and electron microscopy. Biomacromolecules. 2005;**6**:3160-3165. DOI: 10.1021/bm050479t

[122] Dufresne A, Vignon MR. Improvement of starch film performances using cellulose microfibrils. Macromolecules. 1998;**31**: 2693-2696. DOI: 10.1021/ma971532b

[123] Samir A, Alloin F, Paillet M, Dufresne A. Tangling effect in fibrillated cellulose reinforced nanocomposites. Macromolecules. 2004;**37**:4313-4316

[124] Mathew AP, Dufresne A. Morphological investigation of nanocomposites from sorbitol plasticized starch and tunicin whiskers. Biomacromolecules. 2002;**3**:609-617. DOI: 10.1021/bm0101769

[125] Ruiz MM, Cavaille JY, Dufresne A, Graillat C, Gerard J-F. New waterborne epoxy coatings based on cellulose nanofillers. Macromolecular Symposia. 2001;**169**:211-222. DOI: 10.1002/1521-3900(200105)169:1<211::AID-MASY211>3.0.CO;2-H

[126] Favier V, Canova GR, Cavaillé JY, Chanzy H, Dufresne A, Gauthier C. Nanocomposite materials from latex and cellulose whiskers. Polymers for Advanced Technologies. 1995;**6**:351-355. DOI: 10.1002/pat.1995.220060514

[127] Asrofi M, Abral H, Kasim A, Pratoto A, Mahardika M, Hafizulhaq F. Mechanical properties of a water hyacinth nanofiber cellulose reinforced thermoplastic starch bionanocomposite: Effect of ultrasonic vibration during processing. Fibers. 2018;**6**:40. DOI: 10.3390/fib6020040

[128] Bondeson D, Oksman K. Polylactic acid/cellulose whisker nanocomposites modified by polyvinyl alcohol. Composites Part A: Applied Science and Manufacturing. 2007;**38**:2486-2492. DOI: 10.1016/j.compositesa.2007.08.001

[129] Petersson L, Mathew AP, Oksman K. Dispersion and properties of cellulose nanowhiskers and layered silicates in cellulose acetate butyrate nanocomposites. Journal of Applied Polymer Science. 2009;**112**:2001-2009. DOI: 10.1002/app.29661

[130] Magalhaes E, Luiz W, Cao X, Ramires MA, Lucia LA. Novel all-cellulose composite displaying aligned cellulose nanofibers reinforced with cellulose nanocrystals. Tappi Journal. 2011;**10**:19-25

[131] Li Q, Zhou J, Zhang L. Structure and properties of the nanocomposite films of chitosan reinforced with cellulose whiskers. Journal of Polymer Science Part B: Polymer Physics. 2009; **47**:1069-1077. DOI: 10.1002/polb.21711

[132] Azeredo HMC, Mattoso LHC, Avena-Bustillos RJ, Filho GC, Munford ML, Wood D, et al. Nanocellulose reinforced chitosan composite films as affected by nanofiller loading and plasticizer content. Journal of Food Science. 2010;**75** 1-7. DOI: 10.1111/j.1750-3841.2009.01386.x

[133] de Mesquita JP, Donnici CL, Pereira FV. Biobased nanocomposites from layer-by-layer assembly of cellulose nanowhiskers with chitosan. Biomacromolecules. 2010;**11**:473-480. DOI: 10.1021/bm9011985

[134] Lu P, Hsieh Y-L. Cellulose nanocrystal-filled poly(acrylic acid) nanocomposite fibrous membranes. Nanotechnology. 2009;**20**:415604. DOI: 10.1088/0957-4484/20/41/415604

[135] Jean B, Dubreuil F, Heux L, Cousin F. Structural details of cellulose nanocrystals/polyelectrolytes multilayers probed by neutron reflectivity and AFM. Langmuir. 2008; **24**:3452-3458. DOI: 10.1021/la703045f

[136] Podsiadlo P, Choi S-Y, Shim B, Lee J, Cuddihy M, Kotov NA. Molecularly engineered nanocomposites: Layer-by-layer

assembly of cellulose nanocrystals. Biomacromolecules. 2005;**6**:2914-2918. DOI: 10.1021/bm050333u

[137] Chauve G, Heux L, Arouini R, Mazeau K. Cellulose poly(ethylene-co-vinyl acetate) nanocomposites studied by molecular modeling and mechanical spectroscopy. Biomacromolecules. 2005;**6**:2025-2031. DOI: 10.1021/bm0501205

[138] Dufresne A, Kellerhals MB, Witholt B. Transcrystallization in mcl-PHAs/cellulose whiskers composites. Macromolecules. 1999;**32**:7396-7401. DOI: 10.1021/ma990564r

[139] Dubief D, Samain E, Dufresne A. Polysaccharide microcrystals reinforced amorphous poly(β-hydroxyoctanoate) nanocomposite materials. Macromolecules. 1999;**32**:5765-5771. DOI: 10.1021/ma990274a

[140] Pandey JK, Kim CS, Chu WS, Choi WY, Ahn SH, Lee CS. Preparation and structural evaluation of nano reinforced composites from cellulose whiskers of grass and biodegradable polymer matrix. Journal of Composite Materials. 2012;**46**:653-663. DOI: 10.1177/0021998312438174

[141] Fortunati E, Armentano I, Zhou Q, Iannoni A, Saino E, Visai L, et al. Multifunctional bionanocomposite films of poly(lactic acid), cellulose nanocrystals and silver nanoparticles. Carbohydrate Polymers. 2012;**87**: 1596-1605. DOI: 10.1016/j.carbpol.2011.09.066

[142] Hamad WY, Chuanwei M. Nanocomposite biomaterials of nanocrystalline cellulose (NCC) and polylactic acid (PLA). U.S. Patent 8,829,110; 2014

[143] Xiang C, Joo YL, Frey MW. Nanocomposite fibers electrospun from poly(lactic acid)/cellulose nanocrystals. Journal of Biobased Materials and Bioenergy. 2009;**3**:147-155. DOI: 10.1166/jbmb.2009.1016

[144] Salmieri S, Islam F, Khan RA, Hossain FM, Ibrahim HMM, Miao C, et al. Antimicrobial nanocomposite films made of poly(lactic acid)-cellulose nanocrystals (PLA-CNC) in food applications: Part A—Effect of nisin release on the inactivation of *Listeria monocytogenes* in ham. Cellulose. 2014; **21**:1837-1850. DOI: 10.1007/s10570-014-0230-6

[145] Dong H, Strawhecker KE, Snyder JF, Orlicki JA, Reiner RS, Rudie AW. Cellulose nanocrystals as a reinforcing material for electrospun poly(methyl methacrylate) fibers: Formation, properties and nanomechanical characterization. Carbohydrate Polymers. 2012;**87**: 2488-2495. DOI: 10.1016/j.carbpol.2011.11.015

[146] Liu H, Liu D, Yao F, Wu Q. Fabrication and properties of transparent polymethylmethacrylate/cellulose nanocrystals composites. Bioresource Technology. 2010;**101**: 5685-5692. DOI: 10.1016/j.biortech.2010.02.045

[147] Horvath AE, Lindström T, Laine J. On the indirect polyelectrolyte titration of cellulosic fibers. Conditions for charge stoichiometry and comparison with ESCA. Langmuir. 2006;**22**: 824-830. DOI: 10.1021/la052217i

[148] Zhou C, Chu R, Wu R, Wu Q. Electrospun polyethylene oxide/cellulose nanocrystal composite nanofibrous mats with homogeneous and heterogeneous microstructures. Biomacromolecules. 2011;**12**:2617-2625. DOI: 10.1021/bm200401p

[149] Helbert W, Cavaille JY, Dufresne A, Fourier UJ. Thermoplastic nanocomposites filled with wheat straw cellulose whisker. Part 1: Processing and mechanical behavior. Polymer

Composites. 1996;**17**:604-611. DOI: 10.1002/pc.10650

[150] Oksman K, Mathew AP, Bondeson D, Kvien I. Manufacturing process of cellulose whiskers/polylactic acid nanocomposites. Composites Science and Technology. 2006;**66**: 2776-2784. DOI: 10.1016/j.compscitech.2006.03.002

[151] Fortunati E, Puglia D, Monti M, Santulli C, Maniruzzaman M, Kenny JM. Cellulose nanocrystals extracted from okra fibers in PVA nanocomposites. Journal of Applied Polymer Science. 2013;**128**:3220-3230. DOI: 10.1002/app.38524

[152] Alain D, Danièle D, Michel RV. Cellulose microfibrils from potato tuber cells: Processing and characterization of starch-cellulose microfibril composites. Journal of Applied Polymer Science. 2000;**76**:2080-2092. DOI: 10.1002/(SICI)1097-4628(20000628)76:143.0.CO;2-U

[153] Peresin MS, Habibi Y, Vesterinen A, Rojas OJ, Pawlak JJ, Seppa JV. Effect of moisture on electrospun nanofiber composites of poly(vinyl alcohol) and cellulose nanocrystals. Biomacromolecules. 2010; **11**:2471-2477

[154] Li W, Yue J, Liu S. Preparation of nanocrystalline cellulose via ultrasound and its reinforcement capability for poly (vinyl alcohol) composites. Ultrasonics Sonochemistry. 2012;**19**:479-485. DOI: 10.1016/j.ultsonch.2011.11.007

[155] Zoppe JO, Peresin MS, Habibi Y, Venditti RA, Rojas OJ. Reinforcing poly (ε-caprolactone) nanofibers with cellulose nanocrystals. ACS Applied Materials & Interfaces. 2009;**1**: 1996-2004. DOI: 10.1021/am9003705

[156] Habibi Y, Goffin A-L, Schiltz N, Duquesne E, Dubois P, Dufresne A. Bionanocomposites based on poly(ε-caprolactone)-grafted cellulose nanocrystals by ring-opening polymerization. Journal of Materials Chemistry. 2008;**18**:5002. DOI: 10.1039/b809212e

[157] Habibi Y, Dufresne A. Highly filled bionanocomposites from functionalized polysaccharide nanocrystals. Biomacromolecules. 2008;**9**:1974-1980. DOI: 10.1021/bm8001717

[158] Ljungberg N, Bonini C, Bortolussi F, Boisson C, Heux L, Cavaillé. New nanocomposite materials reinforced with cellulose whiskers in atactic polypropylene: Effect of surface and dispersion characteristics. Biomacromolecules. 2005;**6**:2732-2739. DOI: 10.1021/bm050222v

[159] Ljungberg N, Cavaillé JY, Heux L. Nanocomposites of isotactic polypropylene reinforced with rod-like cellulose whiskers. Polymer. 2006;**47**: 6285-6292. DOI: 10.1016/j.polymer.2006.07.013

[160] Rojas OJ, Montero GA, Habibi Y. Electrospun nanocomposites from polystyrene loaded with cellulose nanowhiskers. Journal of Applied Polymer Science. 2009;**113**:927-935. DOI: 10.1002/app.30011

[161] Li S, Gao Y, Bai H, Zhang L, Qu P, Bai L. Preparation and characteristics of polysulfone dialysis composite membranes modified with nanocrystalline cellulose. BioResources. 2011;**6**:1670-1680

[162] Auad ML, Richardson T, Hicks M, Mosiewicki MA, Aranguren MI, Marcovich NE. Shape memory segmented polyurethanes: Dependence of behavior on nanocellulose addition and testing conditions. Polymer International. 2012;**61**:321-327. DOI: 10.1002/pi.3193

[163] Marcovich NE, Auad ML, Bellesi NE, Nutt SR, Aranguren MI.

Cellulose micro/nanocrystals reinforced polyurethane. Journal of Materials Research. 2006;**21**:870-881. DOI: 10.1557/jmr.2006.0105

[164] Pei A, Malho J-M, Ruokolainen J, Zhou Q, Berglund LA. Strong nanocomposite reinforcement effects in polyurethane elastomer with low volume fraction of cellulose nanocrystals. Macromolecules. 2011;**44**: 4422-4427. DOI: 10.1021/ma200318k

[165] Chazeau L, Cavaille JY, Canova G, Dendievel R, Boutherin B. Viscoelastic properties of plasticized PVC reinforced with cellulose whiskers. Journal of Applied Polymer Science. 1999;**71**: 1797-1808. DOI: 10.1002/(SICI)1097-4628(19990314)71:11<1797:AID-APP9>3.0.CO;2-E

[166] Chazeau L, Cavaillé J, Terech P. Mechanical behaviour above Tg of a plasticised PVC reinforced with cellulose whiskers; a SANS structural study. Polymer. 1999;**40**:5333-5344. DOI: 10.1016/S0032-3861(98)00748-4

[167] Chazeau L, Cavaillé JY, Perez J. Plasticized PVC reinforced with cellulose whiskers. II. Plastic behavior. Journal of Polymer Science. Part B: Polymer Physics. 2000;**38**:383-392. DOI: 10.1002/(SICI)1099-0488(20000201)38:3<383::AID-POLB5>3.0.CO;2-Q

[168] Qi H, Cai J, Zhang L, Kuga S. Properties of films composed of cellulose nanowhiskers and a cellulose matrix regenerated from alkali/urea solution. Biomacromolecules. 2009;**10**: 1597-1602. DOI: 10.1021/bm9001975

[169] Ma H, Zhou B, Li HS, Li YQ, Ou SY. Green composite films composed of nanocrystalline cellulose and a cellulose matrix regenerated from functionalized ionic liquid solution. Carbohydrate Polymers. 2011;**84**: 383-389. DOI: 10.1016/j.carbpol.2010.11.050

[170] Wang Y, Cao X, Zhang L. Effects of cellulose whiskers on properties of soy protein thermoplastics. Macromolecular Bioscience. 2006;**6**: 524-531. DOI: 10.1002/mabi.200600034

[171] Cao X, Chen Y, Chang PR, Stumborg M, Huneault MA. Green composites reinforced with hemp nanocrystals in plasticized starch. Journal of Applied Polymer Science. 2008;**109**:3804-3810. DOI: 10.1002/app.28418

[172] Cao X, Chen Y, Chang PR, Muir AD, Falk G. Starch-based nanocomposites reinforced with flax cellulose nanocrystals. Express Polymer Letters. 2008;**2**:502-510. DOI: 10.3144/expresspolymlett.2008.60

[173] Lu Y, Weng L, Cao X. Biocomposites of plasticized starch reinforced with cellulose crystallites from cottonseed linter. Macromolecular Bioscience. 2005;**5**:1101-1107. DOI: 10.1002/mabi.200500094

[174] Saxena A, Ragauskas AJ. Water transmission barrier properties of biodegradable films based on cellulosic whiskers and xylan. Carbohydrate Polymers. 2009;**78**:357-360. DOI: 10.1016/j.carbpol.2009.03.039

[175] Saxena A, Elder TJ, Kenvin J, Ragauskas AJ. High oxygen nanocomposite barrier films based on xylan and nanocrystalline cellulose. Nano-Micro Letters. 2010;**2**:235-241. DOI: 10.3786/nml.v2i4.p235-241

[176] Saxena A, Elder TJ, Ragauskas AJ. Moisture barrier properties of xylan composite films. Carbohydrate Polymers. 2011;**84**:1371-1377. DOI: 10.1016/j.carbpol.2011.01.039